PRELUDE TO THE DUST BOWL

PRELUDE TO
THE DUST BOWL

*Drought in the Nineteenth-Century
Southern Plains*

KEVIN Z. SWEENEY

University of Oklahoma Press : Norman

Library of Congress Cataloging-in-Publication Data

Names: Sweeney, Kevin Z., 1959–
Title: Prelude to the Dust Bowl : drought in the nineteenth-century southern plains /
    Kevin Z. Sweeney.
Description: Norman : University of Oklahoma Press, [2016] | Includes bibliographical
    references and index.
Identifiers: LCCN 2015044574 | ISBN 978-0-8061-5340-7 (hardcover : alk. paper)
Subjects: LCSH: Droughts—Southwestern States—History. | Droughts—United States—
    History. | Southwestern States—History. | Droughts—Social aspects—Southwestern
    States. | Droughts—Economic aspects—Southwestern States. | Dust Bowl Era,
    1931–1939. | Indians of North America—Southwestern States. | Indians of North
    America—Confederate States of America—Government relations. | Indian Removal,
    1813–1903. | Indian Territory.
Classification: LCC QC929.27.S68 S94 2016 | DDC 978/.02—dc23
LC record available at http://lccn.loc.gov/2015044574

1 2 3 4 5 6 7 8 9 10

Contents

Illustrations

FIGURES

MAPS

Acknowledgments

⌒

As with any major endeavor, I owe a debt to many people in the formation of this book. Foremost among them are all the historians of the southern plains who have preceded me and upon whose shoulders I have striven to add to the region's story. The bibliography of this text stands as a tribute to their work. Added to them are the numerous archivists, county secretaries, and librarians who pointed me to treasures within their holdings. A major thanks goes to my professors at Oklahoma State University. The first step in my journey into the role of drought in the southern plains began in the methods class at my alma mater, under the guidance of Richard Rohrs. As I took this topic up as my dissertation, the direction of George Moses and Mike Logan along with that of readers Brad Bays and Neil Hackett gave me organizational insights and helpful critiques of style. My mother and father, Wanda and Bob Partnoy, painstakingly read over each chapter and attempted to instruct me in the mysterious ways of the comma. Lastly, my children, Duncan and Morgan, allowed me to pursue the goal of publication by losing me for hours in archives while on trips, and my wife, Robin, has read and discussed the manuscript ad nauseum, encouraging me when I felt my words were inadequate. Thank you all, and I hope this final product is not an embarrassment.

Portions of chapter 1 were previously published as Kevin Z. Sweeney, "Wither the Fruited Plain: The Long Expedition and the Description of the 'Great American Desert,'" *Great Plains Quarterly* 25, no. 2 (2005): 105–18, and are reprinted with permission.

ACKNOWLEDGMENTS

Portions of chapter 4 were previously published as Kevin Z. Sweeney, "Thirsting for War, Hungering for Peace: Drought, Bison Migrations, and Indian Tribes in the Southern Plains: 1854–1859," *Journal of the West* 41 (Summer 2002): 71–78, and are reprinted with permission.

Earlier versions of chapters 5, 6, and 9 were published in the *West Texas Historical Association Yearbook,* as Kevin Z. Sweeney, "Pandora's Drought: Aridity and the Brazos and Clear Fork Indian Reserves," vol. 79 (2003): 43–55; Kevin Z. Sweeney, "The Evaporating Frontier: Drought and the Northwest Texas Frontier, 1854–1865," vol. 81 (2005): 7–18; and Kevin Z. Sweeney, "Drought in the Heart of Texas, 1854–1865," vol. 84 (2008): 58–73, and are reprinted with permission.

Earlier versions of chapters 7 and 13 were published in the *Chronicles of Oklahoma,* as Kevin Z. Sweeney, "Twixt Scylla and Charybdis: Environmental Pressure on the Choctaw to Ally with the Confederacy," vol. 85 (Spring 2007): 4–33; and Kevin Z. Sweeney, "And the Skies Were Not Cloudy All Day: Drought and the Cherokee Strip Land Run, 1893–97," vol. 81 (Fall/Winter 2003): 436–57, and are reprinted with permission.

Portions of chapter 12 were previously published as Kevin Z. Sweeney, "Wishful Thinking: Attempts to Create Rainfall on the Southern Plains during the 1890s," *Panhandle-Plains Historical Review* 83 (2013): 43–51, and are reprinted with permission.

Introduction

In the summer of 1893, Katharine Lee Bates, a Wellesley College English professor, spent the term teaching at Colorado College in Colorado Springs. Already an accomplished poet and author at thirty-four, Bates joined a group that planned to summit Pike's Peak, whose magnificent dome towered over the town. It must have been an odd scene: delicate women in Victorian dress and frail bespectacled male professors riding a prairie wagon along the rough mountain road, mounting mules when the ruts gave way to a thin trail. When the trail became too narrow even for these beasts of burden, the resolute scholars finished the ascent on foot. The English professor, utterly exhausted, finally topped the 14,000-foot peak, turned a weary gaze east, and was filled with rapture at the breathtaking view. Stunning panoramas of creation can do that to mortals.

Struggling to breathe in the thin air, Bates sat on a rock and scribbled down some lines inspired by the vista. She titled her poem "America the Beautiful," and after its publication in the *Congregationalist* two years later, it quickly became popular as Americans identified with the optimism of its lines. Soon after publication, her poem was set to Samuel Augustus Ward's tune "Materna" and became one of our nation's most beloved anthems. Today her compelling lines are so ingrained in the American mind that one is hard-pressed to read them without recalling the accompanying tune: "O beautiful for spacious skies, / For amber waves of grain, / For purple mountain majesties / Above the fruited plain!"

During the summer of Bates's visit to Colorado, the southern Great Plains experienced a bout with that recurring condition that plagues the region: drought. Just as it was during the depression of the storied Dust Bowl years, the "fruited plain" was a parched and dusty expanse during the economic crisis that closed the nineteenth century, yet we rarely talk about that earlier depression and drought. Instead, if we consider at all the impact of the environment on Americans, we dwell on the infamous brownouts of the 1930s. Still less do we consider the drought of the 1820s or the most severe of all of these, the mid-nineteenth-century drought. There are ample reasons for the dearth of literature on this subject, for the Native people residing on the southern plains and adjoining tallgrass prairie did not keep written records of what they saw and experienced during those dire years. The non-Indians who traversed the region assumed that what they saw was the normal or average condition of the landscape and described it as such. Nonetheless, we can tease out information from their journals and reports and analyze it along with the climate reconstructions that modern technology affords.

Since its emergence in the 1970s, paleoclimatology has blossomed into a multifaceted scientific inquiry that can give us moderns a more complete understanding of the past. All over North America, researchers are boring ice cores, taking tree ring specimens, studying pack rat middens, collecting pollen samples from lake bed sediments, and sharing their findings with one another. Rarely does this information make it to a broader reading public, but the National Oceanic and Atmospheric Administration, formed in 1970, created an entity called the National Climatic Data Center to act as the central nervous system in processing these studies. Through the development of the discipline and its institutions, we as a society have come to comprehend that as Earth's human and animal populations grow rapidly, weather and climatic change have an increasingly harmful impact on the planet's inhabitants and that, in turn, humankind has made its own lasting impression on the environment. Thus, it behooves us to understand the ramifications of past climatic changes and earlier societies' interactions with their terrestrial surroundings.

Focusing on major drought episodes of the nineteenth century in the southern plains and neighboring tallgrass prairies reveals a different understanding of the environmental impact of human activities in the region than we have previously entertained. It is all too easy for a historian living with the conveniences of modernity to overlook the importance of environment on the daily lives of our forefathers. We can remain in our offices typing away on our computers throughout the night under artificial light to meet publication deadlines; with the flick of a switch, we can warm our-selves with electric heat produced by coal plants hundreds of miles away; and we can remain in a region of little water by relying on powerful carbon fuel–driven pumps to extract the life-sustaining liquid from far beneath the soil on which we stand. None of these options were available to those residing in the region prior to the 1890s. Although modern historians have rejected the concept of environmental determinism, and rightly so, in 1899 Paul Vidal de la Blache proposed an alternative in the theory of Possibil-ism. This concept claims that the environment of a given ecosystem places restrictions on the options open to humans, not determining their actions, but altering the possibilities available to them. With this in mind, we can take another look at the events of the first half of the nineteenth century, taking care to account for the impact of climate on the availability of life's necessities in a region that experiences wide fluctuations in rainfall.

During the exploration and settlement of the south-central United States, the cyclical nature of climate was narrowly perceived and often se-verely misunderstood. Certainly, these repetitive climatic patterns and the misperceptions of them were not the only factors affecting the development of the region, but they helped shape the economic, social, and political forces that evolved there. We must always come back to the fact that cli-mate and short-term climate change were pervasive factors underlying all of the others, including relations between Native people and non-Indians, military operations, economic concerns, and even slavery debates. It can be scary to think that many momentous events could have turned out quite differently had they occurred a mere ten years earlier, or if they had been delayed until after the mid-nineteenth-century drought.

Our story begins with an exploratory U.S. military expedition in 1819 and ends with the fin de siècle. During this period, four major droughts bracketed intervening wet trends in the region. The events considered in the following pages occurred for the most part within the modern confines of Texas, Oklahoma, northeastern New Mexico, southeastern Colorado, and southern Kansas—a region without a unifying name, but which the environmental historian Dan Flores has referred to as the Near Southwest.[1] One justification he gives for subsuming so widespread a region under one name is that the area is linked by the flow of major rivers, by history, and by trade. Certainly the region is dominated by grasslands, in particular the short-grass and tallgrass prairies that experienced repeated dry years during the era in question.

Historically, drought has had similar effects on human and animal populations regardless of physical location or culture. When there is little rainfall, wells run dry, springs cease to release water, crops wither, and famine ensues. Human and animal populations then migrate out of the region to find water and food. Those who remain must compete for vital resources, often giving rise to increased violence in the form of crime or warfare. These dangerous attendants to drought have more recently been sidestepped through government subsidies that allow people to remain in a parched environment, as witnessed in the New Deal programs of the Dust Bowl years.

I became acquainted with the climatic moods of the southern plains while growing up in the region. The summer days I spent driving tractors over the sunbaked earth of south-central Oklahoma, often in the open air with only an umbrella canopy to shelter my sweaty head from the sun, inspired my quest to understand humans' relationship to water. During those days I dreamt of rain, not for the benefit of the crops, as I should have, but for the promise of a day off the machine. While breaking the sod, I daydreamed of what life must have been like for the Native people living in this element-tortured environment. Summers in the region can bring temperatures of 116 degrees Fahrenheit, and winters with weeks of freezing temperatures, sometimes dipping below zero, are common. Months might pass with little to no rainfall, but storm clouds can bring dangerous deluges of water to flood the river valleys.

One summer while visiting Arizona, I decided to pilgrimage to Dragoon Springs with my affable older brother. As we exited Interstate 10 and began the climb into the Dragoon Mountains on a dirt road in a two-wheel-drive Chevy Blazer, the reality hit me that this is and was a harsh environment. The only way to survive was to know the location of reliable sources of water. The Butterfield Overland Company understood that problem and negotiated with the Chiricahua Apaches to construct a stage post in the mountain pass, high enough in elevation to reduce the effects of the scorching summer heat. The wise Apache leader Cochise agreed to most of the terms but stipulated that the travel company could not construct the depot within two miles of Dragoon Springs, the only perennial water supply in the region. As my brother and I entered the pass, we located the historical marker, pulled to the side of the road, and stepped out of our modern vehicle into the past. The heat was overwhelming even at that altitude. Thinking we could drive to the springs, we had brought no canteens, but we were sorely mistaken. The only way to see the fabled oasis of water was to hike. Forty minutes of trudging a desert trail made the message clear. With chapped lips and sweat-soaked shirts we finally stumbled to the Dragoon Springs of Apache War fame. A small pool no more than three yards long and two yards wide seeped from the canyon wall. We drank from it thankfully and remarked that men had died fighting over this puddle. The importance of water fully hit home.

PRELUDE TO THE DUST BOWL

ONE

The Great American Desert

Many educated Americans in the first half of the eighteenth century held an opinion that differed greatly from that of Katharine Bates, whose idyllic poem in the second half of the century depicted America's plains as amber waves of grain framed by majestic purple mountains. Through their publications and lectures, they, in turn, convinced others to think of the vast steppe situated between the Mississippi River and the Rocky Mountains as a large wasteland, a barrier to westward expansion.[1] The Long Expedition of 1820 did more than its fair share to promulgate the idea that this region was a great American desert.[2] A team of naturalists who cataloged and collected samples of the animal and plant life they encountered gave the expedition scientific weight. Thomas Say, the mission's zoologist, reported that the group dreaded the journey across "the trackless desert which still separated [them] from the utmost boundary of civilisation," and Dr. Edwin James, the official chronicler of the expedition, stated that the explorers passed "through a barren and desolate region."[3] James claimed that beyond the ninety-sixth meridian, travelers could expect a "wide sandy desert, stretching westward to the base of the Rocky Mountains."[4] This seems incredible to present-day observers, for the ninety-sixth meridian runs near the beautifully green city of Tulsa in humid eastern Oklahoma. Yet the Long Expedition's official report was illustrated by a map labeling today's Great Plains the "Great American Desert." After the accounts and report of the expedition became public, the number of textbook references

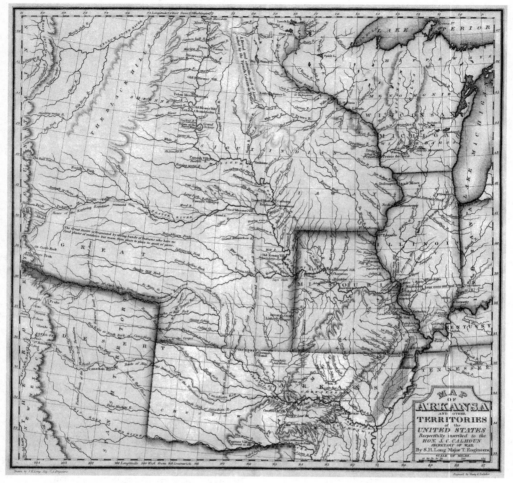

The Stephen H. Long Arkansas Map of 1822, 1500.019. Department of Special Collections and University Archives, McFarlin Library, University of Tulsa.

to the plains as a desert jumped dramatically.[5] Although historians have debated how popular this description became and how long it dominated thought on the region, the Long Expedition will for some time be known for its description of the plains as a Great American Desert.

While modern scholarship has taken an interest in Stephen H. Long's exploration, neither modern nor earlier historians have considered what led the members of the expedition to reach these conclusions. Culture, education, and experience certainly influence how people perceive a region, and these factors must have guided the Long party's portrayal of the southern plains. Environmental factors can also color the way we see a landscape, but it is a common mistake to judge a region's climate only by a brief experience of it. On a mild spring day after a recent rain storm has refreshed the porous soil, and while the winds remain pleasantly calm, the high plains of Texas might seem to a traveler like a pleasant place to live. Yet if the tourist experienced one of the notorious dirt storms with wind gusts of seventy miles per hour blowing tumbleweeds into highline wires, driving small grains of soil through the tiniest crevices of the car and blanketing the dashboard with a fine layer of dust, that traveller would receive a sober awakening. Nature can fool us, and at times it hoodwinked our predecessors as well. Long and his men journeyed through a dry environment at its most arid and assumed incorrectly that this aridity was a constant characteristic of the region.

During the second decade of the nineteenth century, little information was accessible to the educated officer class of the U.S. military concerning the plains between the Mississippi River and the mountains of the West, and what reports did exist were contradictory. The early Louisiana fur traders, though usually tightlipped, could be encouraged to talk about the region with an adequate administration of spirituous liquors; then they tended toward depicting the interior plains as desert-like. But how much of this information was available to the officers of the Long party is unknown.[6]

Lewis and Clark may have been influenced by this desert image. In 1804 Meriwether Lewis wrote to his mother that "from previous information I had been led to believe [the region above the Platte River] was barren, sterile, and sandy." He was pleased to note that he found it quite to the contrary and gave promising comments about much of the plains, with the exception of a stretch along the Missouri River the explorers referred to as the "deserts of America." Clark claimed he "did not think [this region could] ever be settled," while Lewis referred to it as "desert, barren country."[7] Aside from that small section, Lewis and Clark gave glowing reports of the region they traversed. Likewise, President Thomas Jefferson, perhaps indulging in a bit of political spin to justify his highly controversial purchase of the territory, presented a very positive description in his "Official Account of Louisiana," claiming the absence of trees was caused by over-rich soil.[8] Indeed!

Zebulon Pike, whose own findings were first published in 1810, did not share that favorable view. While crossing the central grasslands, he stated that "the vast plains of the western hemisphere may become in time equally celebrated as the sandy deserts of Africa."[9] There were also numerous references in Pike's writings to deserts, aridity, and lack of vegetation.[10] Pike's report influenced Elijah Parish, a Congregationalist preacher and compiler of topographic information, to update his geography text, *A New System of Modern Geography*. In his first four printings, Parish made no mention of a desert but stated in his 1814 revision (after Pike's report) that "between the great rivers, Missouri and Rio Bravo, vast sandy deserts present a dismal prospect; not a tree nor shrub relieves the eye; the salt in the soil forbids vegetation, as in the Tehama of Arabia, and renders the wilds of Louisiana, as cheerless and forlorn, as the deserts of Tartary or Africa."[11] He attributes the description to Pike. It must be added that like many New Englanders, Parish was opposed to expansion for fear that the northeast's political influence would wane with the addition of new states' congressmen to the national legislature. The concept of a desert in the plains could be useful in discouraging the nation's western growth. Regardless of Parish's motives, his text very possibly influenced the educated members of Long's expedition. Parish's text was available in Boston and Newburyport, Massachusetts, as well as Portland, Maine; various New England universities endorsed it.[12]

In 1817 the published notes of John Bradbury and Henry Brackenridge, two English gentlemen who had traveled through the plains, concurred with Pike. They depicted the Missouri River country as "having some resemblance to the Steppes of Tartary, or the Saara's [*sic*] of Africa."[13] It is difficult to believe that Long and his men had read all of these reports, but it is not far-fetched to consider that they were familiar with the traditional view of the area they were preparing to explore.

The background of the primary chroniclers further influenced them to perceive the region's aridity in exaggerated terms. The journalists and scientists were all from the northeastern United States, an area of forests and abundant rainfall. The expedition of 1820 was the first trip into the southern plains region for three of the team's leading members: Stephen Long, the expedition's head who had commanded the Yellowstone expedition the previous year; Edwin James, a Vermont native and the expedition's physician, botanist, and author of its official account; and Philadelphia-born Thomas Say, the team's zoologist.[14]

The bias of these chroniclers' eastern heritage was magnified by the climatic cycle during their lives. The period from 1800 to 1850 was a wetter than usual era for New England. It is no wonder that the scientists believed they were traversing an "inhospitable desert," for the effects of a drought would seem more pronounced to those accustomed to living in a region and period of greater moisture than what the plains experienced in even an average year.[15]

In fact, the members of Long's expedition recorded that they indeed did have preconceived notions of the geographic character of the plains. In 1819, still early on the journey to the source of the Arkansas River, Thomas Say, the expedition's official chronicler at that time, wrote that "you discover numerous indications both in the soil and its animal and vegetable productions, of an approach to the borders of the great Sandy Desert which stretches eastward from the base of the Rocky Mountains."[16] Say accepted the presence of a sandy desert as fact before he had traveled to, or had seen evidence of, a severely arid region.[17] Upon his arrival in 1820 at the winter quarters of Engineer's Cantonment, just north of present-day Omaha, Nebraska, Edwin James became the official recorder of topographic features for the expedition. James wrote that a group of Indians near Council Bluffs laughed at the recklessness of attempting to cross a country that was at that time "so entirely destitute of water and grass that neither ourselves nor our horses could be subsisted while passing it."[18] Captain John R. Bell felt this incident important enough to log in his own expedition journal as well, proving the importance the warning held in persuading the chroniclers that a wasteland awaited, while also revealing that the Natives were aware of the current drought conditions throughout the region.[19] The Natives' annual hunts on the plains familiarized them with the region's climate, and even they considered the conditions of 1820 severe enough to warrant the warning. Accordingly, James, Bell, and the rest of Major Long's group expected a hostile environment of sand and little water.

In addition to the formidable task set before the exploration party, a financial crisis forced Long to undertake the mission without adequate supplies. Economic crisis hit the nation one year prior to Long's planned exploration to locate the source of the Red River. President James Monroe, in an effort to keep the nation financially solvent, reacted to the Panic of 1819 by cutting government spending; in turn, these cutbacks forced the War Department, headed by avowed proslavery expansionist John C. Calhoun, to reduce its budget, though the demand for more results from western exploration was maintained.[20] These pressures on War Department spending were passed on to Major Long. Less funding was available for provisioning the expedition, but the expectations for his excursion had risen. The lack of sufficient supplies coupled with such ambitious goals created a concentration on speed at the expense of thorough scientific research, greatly hindering the party's chances of collecting ample specimens or making accurate observations about the terrain they crossed.[21]

Not only was the approved funding from the War Department inadequate to provision the party, but Long did not even receive all of the funds that were approved. While Long was in the East visiting his new wife, he met with Calhoun, who promised him two thousand dollars. Long was to pick up that money in Saint Louis on his way back to Engineer's Cantonment. He paused in Saint Louis for two weeks, fulfilling obligations of surveying public lands there before progressing further west. Then he waited an additional week at Franklin, Missouri, hoping that the vital financial resources would arrive and further postponing his expedition's departure.[22] Calhoun had sent the promised money rather tardily on April 28, 1820. It took, on average, six weeks for correspondence to reach Saint Louis from Washington; the funds would have arrived in Saint Louis ten days or so after the expedition had departed from the cantonment in today's eastern Nebraska![23] Major Long, obviously aware of the shortage in his party's provisions, attempted to purchase or requisition provisions from Camp Missouri near Council Bluffs. These actions not only proved fruitless but again delayed the expedition.

Western outposts also felt the cut in War Department funding. To compensate for the lack of financial support from the federal government and to supplement their meager stores, soldiers at Camp Missouri planted their own gardens, but a recent flood had destroyed the garrison's small harvest. Even though Major Long carried a note from the honorable secretary of war granting the power to requisition any provisions necessary to fulfill the expedition's orders, the commander of the fort could provide only a few supplies. Fresh mounts, so crucial in overland travel, were a priority, but the commanding officer considered the few horses at the fort vital and would not spare them.

Combined with the attempt to requisition supplies from Camp Missouri, Long's delays in Saint Louis and Franklin proved critical, for they set the party's starting date back a full month. Long planned to begin traveling toward the source of the Arkansas River on May 1, 1820, but the party was delayed until June 6.[24] This delay pressed the necessity of speed upon the military mission to arrive at Fort Smith before the onset of cold weather, which can be brutal on the high plains, while forcing the party to traverse the southern plains during its driest and hottest months.

The supplies listed in Bell's journal reveal the inadequacy of the party's provisions. Compared to the usual rations issued to soldiers during the early nineteenth century, the provisions at Long's disposal could only supply the group with food for thirty days and meat for fourteen days, a woefully inadequate arrangement for a military expedition expecting to be in the field for four months.[25] Given this minimum of provisions, it is apparent that the major intended to trade with Plains Indians for additional foodstuffs, totally discounting the warning of a parched grassland given earlier.[26] Even the amount of trade goods could be expected to supplement the party's insufficient food rations for only three months, and that required meeting Native people on the journey.[27] The meager provisions left a whole month of rations to be desired. The equipment necessary for collecting topographical data and samples was lacking as well.[28] One wonders just how Major Long proposed to meet the many expectations of the War Department while completing this venture. Yet, as commander of the expedition, Long had no alternative but to proceed with the assignment, regardless of the inadequacy of the party's stores.

Unquestionably, Long felt pressured to command a successful military and scientific venture. The Yellowstone Expedition he led in 1819 had been an abysmal failure.[29] It had headed toward the confluence of the Missouri and Yellowstone Rivers to establish an American presence on the British frontier, but the party never made it past Fort Atkinson, near present-day Omaha, Nebraska. More than one hundred expedition men perished from scurvy that winter.[30]

Long did not spend the winter in Fort Atkinson with the rest of the expedition, but instead traveled east to see Calhoun and ask his approval for a renewed effort to explore the plains. The secretary of war obliged Long by issuing written orders for a new mission to explore the Platte River to the central Rocky Mountains and return along the Red River, which had become the boundary with New Spain as a result of the Adams-Onís Treaty of 1819. Given a second chance, Long chose to conduct the expedition as best he could with what supplies he could muster rather than refuse what seemed an impossible task.[31] In an attempt to ensure success on this new assignment, the major pushed his men and animals to their utmost endurance from the very beginning of the mission.[32]

The shortage of supplies became critical in late July, but James mentions the dwindling rations as early as June 26, less than one month after the party had begun the expedition: "Our small stock of bread was by this time so nearly exhausted, that it was thought prudent to reserve the remainder as a last resort, in case of the failure of a supply of game, or other incident."[33] When the expedition arrived at the point where the Arkansas River emerged from the eastern slopes of the Rocky Mountains, Major Long divided his party as ordered on July 4. He sent Captain Bell with half of the force to continue down the Arkansas River, and led the remainder of the group, including the journalist Edwin James, south and west to locate the Red River. Even though each group's demands on the local wildlife for meat were reduced by the smaller number of mouths to feed, that did not lessen the per capita shortage of provisions. By July 29, supplies had become so meager that the major cut dinner rations to one ounce of jerked meat.[34]

The acute shortage of food in an unfamiliar region dictated Long's decision to redirect the party's movement. The possibility of starving in the wilderness was a factor in Long's decision to discontinue the search for the Red River and to instead follow a local ravine in the hopes that it would prove to be one of the Red River's tributaries. Instead, it turned out to be the Canadian River. James, with a penchant for understatement, noted in his journal that the party's "suffering from want of provisions . . . had given [them] a little distaste for prolonging farther than was necessary [their] journey towards the southwest."[35] The absence of any contact with Natives made the prospect of trading for supplies bleak, so Long and his men were forced to rely on their hunting skills to keep themselves alive during their journey across the southern plains. As long as there was game present, this strategy could prove successful, but if the wildlife had migrated from the vicinity of the party's route, they would starve.

During the month of August, Long's party suffered most severely from privation. On August 3, James states that the group was "becoming somewhat impatient on account of thirst, having met with no water which [they] could drink for near twenty-four hours," making their circumstances "extremely unpleasant."[36] It appears James maintained a stiff upper lip in the face of looming disaster. Eight days later, the journalist writes that they "had for some days been almost in a starving condition."[37] By this time the lack of supplies and scarcity of traditional game had reduced Long and his men to eating anything their guns could fell, including badger, owl, and wild horse. Before the trek was over they eagerly consumed turtle and bear meat.[38] Even with the extreme limiting of rations, the party found their provisions of food spent by August 24. It would take twenty more days for the starving men to reach Fort Smith.

Summer months on the southern Great Plains are quite demanding, especially on the endurance of those under the immediate influence of the elements. In August of 1820, when the Long Expedition crossed the grasslands of the Texas Panhandle and Western Indian Territory, James recorded temperatures ranging from 96 degrees to 105 degrees Fahrenheit in the heat of the day. It is to be expected that these travelers suffered from heat stress and dehydration as well as the less severe but certainly uncomfortable sunburn and wind-burn. Their lips would have swollen and cracked. The heat would also have caused the few pools of water that usually remained on the plains to evaporate. While following the riverbed, the party endured days without fresh water, as James explains in his entry for August 22: "It had been only two weeks since the disappearance of running water in the river . . . , but during this time we had suffered much from thirst, and had been constantly tantalized with the expectation of arriving at the spot where the river should emerge from the sand."[39]

The findings of the Long Expedition would have been far more reliable if 1819 and 1820 had been climatically average years. Their journals give a snapshot of the region that proves misleading when considered with meteorological trends over the years. The chroniclers never considered that the conditions they observed on the plains were any different from those usually present there, which was their biggest mistake. By taking for granted that 1819 and 1820 were average years, they excluded the possibility that the terrain they noted in their journals was itself suffering from extreme but temporary water deprivation, just as they were.

To understand climatic conditions that precede recorded weather information for the region, tree ring data can provide the type of information needed to substantiate the occurrence of a drought in areas immediately adjacent to the southern plains. Noted tree ring specialist Merlin Lawson conducted one of the earliest studies on drought in the plains. In order to substantiate the Pike and Long Expeditions' evaluations of the plains as a desert, Lawson gathered data from the Upper South Platte, Upper Rio Grande, Upper Missouri, and Lower North Platte river basins. The results of this study prove quite revealing, especially when considered with the reports of major explorations of the West. The trees studied in the Lower North Platte basin showed a growth of 130 percent in 1803 and 120 percent in 1804, and growths of 85 and 90 percent in the Upper Missouri basin for the same years.[40] This research implies above average rainfall for the North Platte region during these years, and near normal rainfall for the Upper

Missouri River area. The findings correlate with the reports of the Lewis and Clark Expedition, which, by the way, failed to note the existence of a large desert region east of the Rocky Mountains. Conversely, the tree ring data show an extremely dry year for 1806 in the Upper South Platte and Upper Rio Grande river basins, with tree growths of 50 and 40 percent respectively.[41] The presence of such severe drought places the report of Zebulon Pike in a more proper perspective, for these are the regions he traveled in 1806.

Lawson's dendrochronological data also substantiate the occurrence of a drought in the southern Great Plains from 1818 to 1820. The Lower North Platte basin, the route of travel for Long and his men during June of 1820, underwent a mild but prolonged drought with tree rings expanding 80, 70, and 80 percent of their average growth for the three years in question.[42] The Upper Rio Grande basin experienced conditions much more harsh, showing tree ring growths of 60 percent for 1818, 40 percent for 1819, and 80 percent for 1820.[43] The region leeward of this basin could expect similar drought conditions due to the prevailing winds, which carry what moisture there is east over the Sangre de Cristo Mountains. As the air is forced up over the mountains, it cools, causing the moisture to condense, thus giving the slopes of these peaks enough rainfall to sustain a conifer forest. If the tree ring data from these trees reveal a drought, then it is probable that the plains downwind of them experienced a drought as well. Notably, this is the region the Long Expedition traversed from July 27 into early August. James describes the results of this extended period of little rainfall in the area: "we were still passing through a barren and desolate region affording no game, and nearly destitute of wood and water."[44] It is evident that the region was showing the telltale signs of acute water shortage to the point that even animals could not survive in the forbidding environment.

Although Lawson did not find the drought conditions derived from this study severe enough in themselves to justify Pike's or Long's portrayal of the southern plains as a desert, his research did trigger further studies of his own as well as some by other scientists.[45] Subsequent research has found evidence of a severe drought in central Oklahoma during 1819. Samples were remarkably similar between locations in Payne and Johnston counties in north-central and south-central Oklahoma, validating widespread arid conditions for that year.[46]

In a later publication, Lawson and coauthor Charles Stockton support the conclusion that a drought did engulf the southern plains in 1806 and

1820.[47] The scientists tested the Palmer Drought Severity Index (PDSI) against the climatological records available for the American plains from 1931 to 1962 and found these regional reconstructions to be surprisingly accurate; however, the tests tended to underestimate extreme conditions.[48] Using the Palmer Index to reconstruct past climates, Lawson and Stockton contend that the area observed by Long's party was in the grip of a drought more severe than the Dust Bowl of the 1930s![49]

Stockton and the University of Arizona's David Meko substantiated this study. During the summers of 1980 and 1981, they collected tree ring samples from sites in Quanah, Texas, near southwestern Oklahoma; in the Arbuckle Mountain region of south-central Oklahoma; and around Lake Eufala in eastern Oklahoma. They successfully tested their method for calibrating the tree rings against weather records from 1933–1977 and found that the 1930s drought was milder than the drought of the 1820s.[50] Such information substantiates the recurring nature of drought in the southern plains.

Another study sponsored by NOAA in concert with the National Geophysical Data Center (NGDC) points to the occurrence of a severe drought in the present-day five-state area of Colorado, New Mexico, Oklahoma, Texas, and Kansas during 1820.[51] Cores extracted from tree samples in the Wichita Mountains, along the Canadian River, on the slope of the Capulin volcano, and in southeastern Colorado tend to corroborate this data and render it quite reliable. NOAA and the NGDC also utilize the Palmer Drought Severity Index to estimate the level of drought experienced by regions where tree ring evidence is not available. The PDSI plots show a year of drought in 1818 followed by increasingly dry years in 1819 and 1820 in the southern and middle plains. Each successive year of low rainfall increases the severity of the drought on the landscape, causing the complete denudation of vegetative cover from the soil in some areas.

In 1998, two dendrochronologists from the University of Arizona, Connie Woodhouse and Jonathan Overpeck, conducted an investigation into past droughts on the central plains. They collected evidence from all available sources of paleoclimatological data, including historical documents, tree rings, archaeological artifacts, lake bed sediment, and geomorphic evidence, to assess the frequency and severity of droughts on the nation's grasslands. These authors found that the southern plains experienced a prolonged severe drought from 1818 to 1822, roughly equivalent in severity to the extreme drought on the southern plains between 1953 and 1956, which was shorter than the 1930s drought but brought with it a more acute short-age of moisture.[52] A 1999 study conducted by Edward Cook, David Meko, David Stahle, and Malcolm Cleaveland concludes that the drought of the 1930s was more severe for the nation *as a whole* than any droughts that have occurred in the past three hundred years, but this does not contradict the earlier studies, which found that the drought of 1818–1822 was more severe than the Dust Bowl era for the southern plains.[53] In fact, the authors of the 1999 study mention that local droughts such as those occurring on the southern plains in the 1820s, 1850s, and 1950s may have proved more severe for their specific locations.

Descriptions of aeolian activity further substantiate the presence of a severe drought on the plains during the Long Expedition's journey. Cer-tain areas adjacent to plains rivers are susceptible to dune formation from blowing sands during extremely dry periods, usually on the eastern banks. During years of average or above average rainfall, these dunes are stabilized by vegetative cover, but during times of extreme aridity, the plants die off and expose the unanchored sand particles to the power of the wind. James described the resulting dunes to the north of the dry Canadian riverbed in present-day northeastern New Mexico: "Extensive tracts of loose sand, so destitute of plants and so fine as to be driven by the winds, occur in every part of the saline sandstone formation southwest of the Arkansa [*sic*]."[54] From James's description we can see that the Canadian River dunes had been destabilized by the prolonged dry trend and were the source material for sand storms when the wind picked up.

Daniel Muhs and Vance Holliday conducted another study of dune ac-tivity in the Chihuahuan Desert, the northern and central Great Plains, and the Colorado Desert of southeastern California. They found that dune activity increases in western North America when precipitation is exceeded by potential evapotranspiration, which increases as a result of wind, sunny days, and heat.[55] Upon examining modern aerial photographs of the Ca-nadian River in 1995, Muhs and Holliday were not able to locate many instances of active dunes. Dune conditions at that time were nowhere near as active as those described repeatedly by James and Long, indicating that the region was experiencing a much more severe shortage of rainfall in 1820 than in 1995.

Aeolian processes were no doubt active during the drought from 1818 to 1822 along the Canadian River. Conditions of low precipitation along with high temperatures decreased the amount of vegetative cover on the sand

and left the riverbed dry. As the surface air heated, it began to expand and rise, creating a vacuum and increasing the winds. These wind gusts picked up the riverbed's sand particles and blew them to the northern banks of the Canadian River, adding them to the dunes that were now devoid of plant cover. In the words of James, "The drifting of sand occasioned much annoyance. The heat of the atmosphere became more intolerable, on account of the showers of burning sand driven against us, with such force as to penetrate every part of our dress, and proving so afflictive to our eyes, that it was with the greatest difficulty we could see to guide our horses."[56] The drought punished the Long party on the part of their journey when they were most vulnerable. As the expedition moved out of the mountains, their supplies grew shorter and the heat from the summer grew more intense, making their journey a memorable one of deprivation and suffering.

James's journal entries contain the strongest evidence for the possible productivity of the plains. It is ironic that the manuscript so often maligned for labeling the plains a "desert" also contains so many references to fertile soils, rich grasses, and abundant wildlife. On July 27, 1820, while the expedition was in the southeast corner of today's Colorado, James described the party's "surprise . . . to witness an aspect of unwonted verdure and freshness in the grasses and other plants of the plains."[57] It was common in the early 1800s to associate soil fertilities with the plants growing on them. Trees, especially, could identify productive soils. James's training in these studies is evident in several of his entries as the party crossed today's Texas Panhandle. James wrote in his journal on August 14, "the occurrence of elm . . . not to be met with in a desert of sand, give[s] us the pleasing assurance of a change we have long been expecting to see in the aspect of the country."[58] The naturalist also notes "the occurrence of the black walnut, for the first time since [they] left the Missouri River," while just twenty miles east of the one hundredth meridian. According to the chronicler, the presence of the black walnut "indicates a soil somewhat adapted to the purpose of agriculture."[59] These two remarks are interspersed through several references to "sterile and sandy" terrain, the "barrenness of the soil," and "extensive tracts of loose sand."[60] It is apparent that James considered much of the land he saw worthless to the future of his nation, and yet, in his more optimistic references, he describes other areas that would be of definite value to his countrymen.

Later in his journal, James qualifies these promising descriptions of the high plains. The main impediments to settling this region, according to the chronicler, are the dense root systems of the grasses and the absence of moisture and timber. On the evening of August 19, he summed up his opinion of the high plains: "The elevated plains we found covered with a plenteous but close-fed crop of grasses. . . . The luxuriance and fineness of the grasses, as well as the astonishing number and good condition of the herbivorous animals of this region, clearly indicate its value for the purpose of pasturage."[61] James continues to restrict the potential of the grasslands by writing that "the soil of the more fertile plains is penetrated with such numbers of [strong roots] as to present more resistance to the plough than the oldest cultivated pastures."[62] This assessment would prove to be correct, but hardly a reason to discourage settlement of the plains. Just one week before the party entered the welcomed confines of Fort Smith, the journalist noted that the soil west of the falls of the Canadian River near the ninety-sixth meridian was "in some places fertile, [but] the want of timber, of navigable streams, and of water for the necessities of life, render it an unfit residence for any but a nomad population."[63]

The influence of the geographic literature on the plains, the lack of supplies, and the summer heat did not inhibit James from writing of what he saw, but it should be noted that the journalist is not as negative about the area to the west of the one hundredth meridian in his original manuscript as he is in the published version.[64] Seemingly, his memories of hunger and thirst colored his final report. Even given all these parameters, James's findings were accurate in describing the terrain for the specific time in which he was crossing it. The effects of the previous years' drought left little moisture in the rivers or streams and reduced the grass cover of the region, exposing sand and driving much of the wildlife to other areas in search of better grazing.

All references to "luxuriant grasses" aside, James and Long succeeded in convincing the government, and the public, that the Great Plains were truly the Great American Desert. James's journal entry for September 6 provides a most telling excerpt: "Speaking of the occurrence of a peculiar bed of rocks crossing the river [the falls of the Canadian River, near the ninety-sixth meridian], . . . when the traveler arrives at this point, he has little to expect beyond, but sandy wastes and thirsty inhospitable steppes."[65] Later in the same paragraph, James adds, "Beyond [fifty or sixty miles above the falls] commences the wide sandy desert, stretching westward to the base of the Rocky Mountains." In his report to Calhoun, Major Long described the region as "almost wholly unfit for cultivation, and of course uninhabitable by a people depending upon agriculture for their subsistence."[66] The personal testimonies of the chroniclers concerning the worthlessness of the land for settlement profoundly impressed the popular perception of the region. The public considered the Long Expedition to be a scientific exploration. By endorsing the idea of the plains as uncultivable, the expedition corroborated the earlier findings of Pike and entrenched in the American mind the concept of a Great American Desert.[67]

A prolific historian of the southern plains, James Malin, recognized the early nineteenth-century explorers' ambiguous usage of the term "desert" in *The Grasslands of North America* to describe areas carpeted with grass.[68] James and Long use the term that way as well, but if one considers their era's definition of the term, the usage is not so perplexing. Noah Webster's *An American Dictionary of the English Language,* published in 1828, defines "desert" as "an uninhabitable tract of land; a region in its natural state; a wilderness; a solitude; particularly, a vast sandy plain, as the deserts of Arabia and Africa. But the word may be applied to an uninhabitable country covered with wood."[69] It seems as if the word "desert" indicated aspects of the unhibabitability of a region rather than specifically denoting an ecosystem of little rainfall. The explorers obviously intended to utilize many of these meanings when depicting the terrain. Certainly the phrase "sandy waste" denotes a dry region in tune with a more contemporary meaning of "desert," while a phrase such as "barren waste" would describe the absence of trees and humans. There is no doubt that the phrase "Great American Desert" was intended by the Long Expedition to include meanings of dry, uninhabited, and treeless, all of which were accurate for the plains in a time of drought.

The journalists of the Long Expedition were influenced to view the Great Plains as unfit for settlement even though there was evidence that the region was habitable. The literature and myth of their time stressed the barrenness of a sandy desert east of the Rockies and guided their perception of the landscape before they even set foot on it. Shortages of food and water forced the party to endure severe privation along their journey, which resulted in memories of dry, "sterile soil." The effects of drought upon the region they traveled must be recognized as pivotal in persuading the journalists that the area was too arid for cultivation, even as far east as the ninety-sixth meridian. As plants died from lack of moisture, they exposed the ground to the effects of the wind and evaporation, giving the appearance of a desert. James and Long took this view of the plains back to Washington with them, and their report promoted the concept of a withered plain to officials back east. Surely, Katharine Bates would have been disappointed to learn that the magnificent Pike's Peak, from which she was inspired to poetry, took its name from an explorer who described her "fruited plain" as a desert of sand. In the span of eight decades, the opinion of North America's Great Plains had changed drastically, but Pike's and Long's assessment would hold for almost another forty years, influencing government policy along the way.

TWO

Long's Report and Indian Removal

John Bell's half of the Long Expedition arrived at Belle Point, a prominent bluff overlooking the Arkansas River, on September 9, 1820, after ninety-three days of trekking through the plains. This strategic location at the confluence of the Poteau and Arkansas Rivers had become the site of Fort Smith in 1817. Long's group of eleven gaunt and exhausted men entered the three-year-old fort four days later, and the reunited officers spent the next day writing reports to Secretary Calhoun. The travelers remained at the post only a few days, recovering from their ordeal and exchanging stories while eating their fill of the fort's food. On September 19 Bell and two others departed to return to their homes back east, followed the next day by Edwin James and a small group that included Stephen Kearny (later to become famous for his command during the Mexican-American War and as governor of California). James and Kearny decided to make a side journey to the hot springs of central Arkansas Territory to enjoy its recuperative effects on their tired bodies. Stephen Long, Thomas Say, Titian Peale, and six others left Fort Smith on September 21, eventually catching up with Bell's group before arriving at Cape Girardeau, Missouri.

James continued to write in his journal as the group made its way to the Mississippi River. On their first night out of Fort Smith, as evening was approaching, his party came upon a small cabin whose owner told them they would receive better accommodations in a house up the trace. They remounted and rode on to "Squire Billingsby's place," where they were treated with southern hospitality. After Mr. Billingsby offered the travelers some honeycomb from one of his beehives during dinner, the Billingsby children gave up their beds for the guests and slept at a neighbor's house. As James recounts, after an "unquiet and sleepless night" in such soft beds, the travelers woke with a "painful feeling of soreness in our bones, so great a change had the hunter's life produced upon our habits."[1] By October 12 the various parties had met at Cape Girardeau and were pleased to find the steamer *Western Traveler* docked at the pier. As a fitting end to their journey through the central plains and frontier, most of the members of the expedition were suffering from "intermittent fevers," probably malaria contracted in the swamps near Fort Smith or Little Rock. Bell and Long completed their trip to Washington, D.C., while Thomas Say and Titian Peale arrived at Philadelphia by January 1821 with what remained of their specimens. James had more difficulties traveling than the others. He had to borrow the funds to pay for his passage back east, and his trunk with all of his money in it was stolen in Maryland. Although he did recover most of his property, the money was never returned and he had to borrow yet more to finish his journey. James finally arrived in the fall of 1821, bitter towards Long because of the hurried manner in which the party traversed the plains.[2]

While in Washington, Stephen Long composed his formal report, and Bell put the finishing touches to his journal for submission to Calhoun. The secretary of war agreed to provide a stipend and offices to Long, Say, and James as they prepared their briefs. As Long continued to fight off bouts of illness, and as Thomas Say fulfilled his duties as a natural history professor at the University of Pennsylvania and curator of the American Philosophical Society, James sifted through the notes of Say and Long and his own journal to conclude the *Account of an Expedition from Pittsburgh to the Rocky Mountains*. The account was published early in 1823 and appeared in bookshops the same year.[3] Expedition members enjoyed a favorable reaction to their scientific work, the collection of flora and fauna samples, and the maps they created. Henry Schoolcraft and the *Niles' Weekly Register* praised the efforts of the exploration of a relatively unknown region of the United States, and the *North American Review* was only critical of the federal government's tightfisted funding of the expedition, expressing disappointment that more wasn't accomplished.[4]

Long's and James's views of the region became widely accepted by the scientific community. In 1823, Edward Everett presented his analysis of the account in the *North American Review*. Everett claimed that Long's party was more qualified to conduct an extensive scientific exploration of the West than Lewis and Clark, and he saw the expedition's work as a valuable addition to scientific knowledge.[5] Jedidiah Morse's *New Universal Gazetteer,* published in 1823, consulted Long's official expedition report and cited James on occasion; the *New Universal Gazetteer* is now considered the most current geographical dictionary of its time.[6] In 1825, the *North American Review* ran an article covering Long's exploration of the Red River of the North, titled "Major Long's Second Expedition." Although it was not concerned with Long's journey across the plains, this essay praised his "zeal and industry" in performing "the expedition to the Rocky Mountains with so much credit to the persons employed, and advantage to the cause of science."[7]

At the time, there were other published articles that used the "desert" characterization to describe the plains. In 1823, Benjamin Silliman, America's first professor of natural philosophy at Yale University, included a description of the Long Expedition's findings in the *American Journal of Science and Arts:* "We perused with no small regret the account of the vast sandy desert, which for the distance of five hundred miles from the feet of the Rocky Mountains, presents a frightful waste, scarcely less formidable to men and animals than the deserts of Zahara [*sic*]."[8] The *Niles' Weekly Register* printed a paragraph titled the "American Desert" that referenced the expedition: "There is an extensive desert in the territory of the United States, west of the Mississippi, which is described in Long's Expedition to the Rocky Mountains. . . . All the elevated surface is barren desert, covered with sand, gravel, pebbles, &c. There are a few plants but nothing like a tree to be seen on these desolate plains, and seldom is a living creature to be met with."[9]

The phrase "Great American Desert" can be found on the maps displayed in various atlases and books from 1823 through 1840. Thomas G. Bradford's *A Comprehensive Atlas, Geographical, Historical and Commercial,* published in 1835, places the so-called desert almost evenly astride the one hundredth meridian, stretching from 97 to 106 degrees longitude. Other atlases place the desert anywhere between the ninety-eighth and one hundredth meridians to the Rocky Mountains.[10] Though there was little uniformity in the understood boundaries of the interior desert, it quickly becomes discernable that James's and Long's reports effectively seared the notion of a Great American Desert on the American mind. Moreover, that idea of an interior desert influenced the destination Congress would choose for relocating the eastern Indian tribes.

The idea of Indian removal can be traced back to Thomas Jefferson. In the Georgia Compact of 1802, the southeastern state agreed to relinquish its western lands in return for a payment of $1,250,000 and the promise that the federal government would terminate tribal rights to their lands within the state boundaries as soon as it could be affected in a peaceable manner. Shortly after the conclusion of the Louisiana Purchase, Jefferson set about drafting an amendment proposal that would give Congress the power to relocate the Native people of the eastern woodlands to the trans-Mississippi region.[11] Although this proposal never became law, several hundred Shawnee and Delaware moved from the Ohio Territory to the eastern Canadian River in present day Oklahoma.

Cherokee relocation to the trans-Mississippi West began even earlier. Prominent war leader The Bowl, though of a mixed-blood Cherokee-Scottish ancestry, held an undying hatred for whites after North Carolina settlers killed his father. In 1794 members of The Bowl's band discovered a party of non-Indian traders including six men, four women, three children, and twenty slaves sailing down the Tennessee River through Cherokee tribal territory. The merchants dispensed alcoholic drink freely, then took advantage of the Cherokees in trade. After attempting to talk the traders into returning his peoples' goods, The Bowl led an attack, killing the men of the group. Worried about the reaction among his fellow tribal members who were trying to abide by treaty conditions with the non-Indian people near them, The Bowl decided to take the surviving emigrants and their boats and descend down the Tennessee, Ohio, and Mississippi Rivers to the mouth of the Saint Francis River south of Memphis in present-day Arkansas, where he and his followers disembarked and allowed the remaining non-Indians to continue on to New Orleans. In 1810 the Cherokee leader Tahlonteskee, frustrated by non-Indian encroachment on tribal lands and convinced of the futility of armed resistance, left his homeland with close to three hundred followers, several hundred head of livestock, and some slaves, and traveled to join The Bowl's settlement. This Cherokee community inevitably attracted other members of the tribe and by 1815 around three thousand Cherokee had gathered on the rivers of north-central Arkansas.

These migrations impacted other people as well, specifically the previous occupants of the lower Arkansas region, the Quapaws and Osages. The Quapaws, or Akansea as they were known by their Algonquin-speaking neighbors, gave their name to Arkansas Territory, but smallpox and other diseases devastated the tribal population. The Osages controlled most of northwestern Arkansas, southwestern Missouri, southeastern Kansas, and northeastern Oklahoma, and they viewed the incursion of other Native groups into their hunting domains as an act of war. The Osages retaliated, striking at any Cherokee hunting parties in their lands. U.S. officials tried to negotiate a peaceful settlement to these hostilities, which threatened to undermine the whole Indian removal policy. In the resulting treaty, known as Lovely's Purchase, the Cherokees gained most of northeastern Arkansas and the remaining land west to the Verdigris River. In 1817 another three thousand Cherokees migrated to the region, and with the added pressure of non-Indian people emigrating to the area, the Cherokees began hunting in Osage lands west of the Verdigris. The Cherokees' repeated trespassing on Osage territory caused a return of hostilities between the tribes and culminated in the Battle of Claremore Mound, in which a Cherokee war party killed eighty-three Osages and captured more than one hundred women and children. In an effort to contain the Indian attacks, the U.S. military constructed Fort Smith on Belle Point in 1817 and Fort Gibson on the Grand River in 1824. One year later with the Treaty of Saint Louis, the Osages relinquished their claim to their lands in present-day Oklahoma and accepted a smaller land claim in Kansas, opening up the land in northeastern Oklahoma to the Cherokees and allowing the removal policy to proceed uninhibited.

Over the next few years, even though Indian settlers continued to emigrate from their lands east of the Mississippi River, the plan to remove the eastern tribes stalled in Washington, as it was tied to the growing controversy over the extension of slavery. Supporters of the "peculiar institution" were acutely aware that relocating the proslavery Five Tribes to the region west of the Mississippi River would extend the plantation lifestyle further west. Once the Missouri Compromise became law in 1820 and created a boundary between slave and free territories in the Louisiana Purchase, the possible land claims for these slaveholding tribes was diminished to the area between the Missouri Compromise line of parallel 36°30" north and the Red River.

Amidst all of the controversies over removal and the extension of slavery, Christian missionaries worked to spread both the Christian gospel and white culture to the Native people. Isaac McCoy, a Baptist missionary working with the Pottawatomie Indians in Michigan Territory, was a key figures in promoting removal of the eastern tribes to the trans-Mississippi West. After he was born in Uniontown, Pennsylvania, in 1784, his family moved to Kentucky where his father pastored a small country church. The family soon moved across the Ohio River into Indiana Territory as his father preached at various rural churches. It was here in the recently settled region of Indiana that Isaac had a religious experience and professed Christ at the age of sixteen. At the age of nineteen, he married Christiana Polk, a cousin of James K. Polk, who became president of the United States in 1845. Christiana's mother had been kidnapped by the Ottawa Indians early in her life, and her mother's tales inspired Christiana's yearning to conduct mission work among the Indians of the Old Northwest.[12] In 1808 the elders of the Silver Creek Baptist Church recommended young McCoy for ordination to preach, but it wasn't until two years later that Isaac assumed the pulpit of Maria Creek Church in Clark County, Indiana Territory. After eight years of pastoring, McCoy embarked on his calling as a missionary. He and Christiana made no differentiation between foreign and home missions and called for Baptist missions to minister to the nearby Indians. McCoy's evangelical stance often led to conflicts with local mission boards, which were not as sympathetic to the plight of Natives as the McCoys.

At the same time, the Restoration Movement called for the unification of the various Christian denominations into a single church and was proving extremely popular in the frontier regions. The increasingly populated frontier did not have enough preachers for every community. Thus, settlements had to make do with one minister for the various denominations, if they were lucky enough to have one. Another characteristic of Christianity during the era was a growing concern with the idea of Christ's second coming, a concern that influenced church members to focus more on their own self-education than on ministering to other groups. The concern with Christ's second coming added to the church's rejection of denominational organizations as not being characteristic of the New Testament church, which in turn reduced financial support for mission boards in general. As a result, many churches flatly rejected organized mission work on the frontier.

Despite the lack of support from other churches, Isaac and Christiana pursued their goal to work with Native peoples and found support in the Baptist Missionary Board in New England. In 1817 the McCoys earned an appointment to minister to the Kickapoo and Miami Indians of the upper Wabash Valley and established the Raccoon Creek Mission the next year. Within two years, McCoy was intent on opening another mission among the Miamis. On a trip to Fort Wayne, McCoy saw for himself the Natives' widespread alcohol abuse and subsequent violence, induced by the apportionment of annuities, and became convinced that his efforts were needed there.

As a result of these experiences, McCoy came to believe that all efforts to "civilize" the Indians were unproductive unless they could be removed from the negative effects of living in contact with whites. The irony is inescapable: McCoy felt that the best chance to acculturate the Natives to Anglo-American mores would first require separating the Indians from whites. In McCoy's eyes, the influence of whiskey peddlers was most damaging to the progress of the aborigines, and it became his personal mission to work for the relocation of his flock to a suitable site for their development. In 1824 McCoy traveled to Washington in the hope of pitching his Indian relocation idea to the president. Although unable to gain an audience with John Quincy Adams, McCoy did visit with Secretary of War Calhoun, who strongly supported the idea. Together they worked relentlessly for Indian removal to a suitable area west of the Mississippi River.

Finally, in June of 1828, after four years of lobbying, Congress approved a reconnaissance trip, with McCoy taking leaders from the Pottawatomie, the Chickasaw, the Choctaw, and the Ottawa Nations to investigate possible sites for their relocation. On August 20 McCoy and ten others left Saint Louis without the Chickasaw and Choctaw delegates. Even before the trip got underway, on July 6, while he was waiting not so patiently for the Chickasaw and Choctaw representatives, he wrote in his journal that "we are limited to the regions west of Arkansas Territory, and Missouri state. Should the inhospitableness of that country deny them a place there, they will be left destitute."[13] He also discussed the "supposed scarcity of water in the prairies" as a concern, which may have impressed him to keep his group near the Missouri and Arkansas borders, for he never escorted an Indian contingent further than fifty miles west of these lines.[14]

As it happens, the Baptist missionary had actually met Stephen H. Long. In his essay "History of the Baptist Indian Missions," McCoy proudly notes the arrival of Long and his second expedition at the Baptist mission on the St. Joseph River in 1823 and quotes the expedition's official journal in describing the progress made toward building a school and clearing fields.[15] In his *Practicability of Indian Reform*, McCoy also cites James as he describes the desert west of the Mississippi River: "Along the vast chain of the snow topped Andes, or Rocky Mountains, nature has spread on each side a barren desert, of irreclaimable sterility. To what extent this sandy desert spreads to the west of those mountains, and what exceptions to its barrenness may occur, we have not the means of knowing. Dr. James allows it an average width on the east side of the mountains of between 500 and 600 miles [around the ninety-sixth meridian]."[16] Curiously, it is toward this region that McCoy wished to relocate the Indians.

McCoy did attempt to justify choosing such an unpromising area for Native relocation. First of all, there was a strip of suitable farmland between the white settlements in Missouri and Arkansas and the ninety-sixth meridian. Jedidiah Morse wrote in his "Report to the Secretary of War" that this area "is said to be some of the finest lands in the Arkansas Territory."[17] Also, the location of the so-called desert immediately to the west of the proposed reservation would preclude any white settlement to the west of the relocated tribes and prevent another encirclement of Native groups.[18] McCoy described at length the undesirability of the area: "The vast region is not termed a desert, merely on account of the almost, or entire, absence of timber, but chiefly because the soil itself is of a quality that it cannot be rendered productive by the industry of man. No portion of our territories furnish so few inducements to civilized man to seek in it a dwelling place, as this under consideration."[19] McCoy seemed to relish relating the plains' shortcomings to his readers even though he had not traveled there. In fact, he shunned Morse's proposal to relocate the tribes in what was considered better country in the Northwest Territory, now Wisconsin and Minnesota, because that land would be desired by white settlers who would force the tribes to move yet again.[20] Finally, McCoy claims that "good grazing country" is the best environment for "people in their transition from hunter to the civilized state."[21]

There is much turbidity in McCoy's writings on the influence of the "Great American Desert" concept as a factor in his removal proposal. There can be no doubt that he was privy to Long's and James's writings and that the desert appealed to his desire to remove the Natives from the negative influences of white culture. McCoy was more interested in making this arrangement permanent than he was in many other aspects of the plan for removal.[22] He was aware of the previous attempts to gain homelands for individual tribes and was committed to procuring a residence that would never be tampered with: thus, the importance of having a desert to the rear of the new tribal lands.

Jeremiah Evarts opposed Indian removal as vehemently as McCoy supported it. The Yale-educated editor of the orthodox Congregationalist journal *Panoplist* visited the southern Indians, particularly the Cherokees, on several occasions, and came to believe that if left to grow and prosper, these people could become "civilized." His "Essays on the Present Crisis in the Condition of the American Indians," published in the *National Intelligencer* in 1829, attacked the state and federal governments for their treatment of Natives.

Evarts was also aware of Long's writings. In his "Memorial of Citizens of Massachusetts," Evarts states, "If the country west of the territory of Arkansas is correctly described by Major Long, an authorized agent of the government, it is uninhabitable."[23] Evarts's ready belief of Long's conclusions conveys the scientific community's acceptance of the expedition's findings as an authoritative contribution to the field. Evarts emphasized Long's credentials and government sponsorship to support the conclusion that the region is not fit for human habitation, including Indians.

In other writings, Evarts echoes the Long Expedition's findings. In his "Memorial of the Prudential Committee American Board of Missions" address to the Senate, Evarts claims that "the western side of this territory [designated for Indian colonization] will be an illimitable desert."[24] He later evaluates the terrain of the proposed site more deeply: "Though considerable uncertainty prevails on the subject, yet it seems admitted by all, that the far greater proportion of the contemplated new residence for the Indians, (probably four-fifths of the whole), is an immense prairie, nearly destitute of wood, deprived of running water four or five months of the year."[25] If such a description were accurate, the location would be obviously unfit for any people attempting to establish an agricultural society. Still later in the essay, he leaves no doubt as to what area he describes with the term "prairie." When discussing the proposed region, Evarts claims that the tribes would be caught between the "pressing white population on the east, and a boundless prairie, . . . often called a desert" on the west.[26]

The reference to Long is obvious. The descriptions of the plains as a region of little moisture and inhospitable terrain is constant. For the same reasons that McCoy considered the area perfect for Indian relocation, Evarts thought it unsatisfactory. The people he knew as Indians, the Cherokees, were agriculturalists just like the white population, not hunters. In Evarts's estimation, Natives relocated to such a barren area would become despondent and resort to government subsidies for support. For these reasons, Evarts contended that the Indians were better off where they were, in a state of self-sufficiency and of little cost to the government.

It appears that some members of the Eastern Band of Cherokee Indians were also informed on the notion that the terrain to the west of the Mississippi River was a desert. In an address to the U.S. Senate in 1824, the Cherokee spokesman voiced concern over the economic options open to them if they moved to the proposed area, relying on a description of the region which strongly resembled the language in James's expeditionary account: "removal to the barren waste bordering the Rocky Mountains, where water and timber are scarcely to be seen, could be for no other object or inducement than to pursue the buffalo, and to wage war with the uncultivated Indians of that hemisphere."[27] According to the Cherokee delegation, relocation to such an area would be a step backwards from civilization, negating the justification for removal.

One of the strongest arguments for removing the tribes to the West was the threat posed by keeping a possibly hostile population within the United States and within close proximity of Mexico and British Canada. In the wake of the War of 1812, many congressmen were concerned about national security, and in 1817, a committee report on the "Exchange of Lands with the Indians" circulated the idea that Indian removal was necessary to preserve national defense.[28] Senator Thomas Reed of Mississippi reintroduced the notion in 1827, reminding his peers of how close the Choctaw and Chickasaw tribes had come to joining Tecumseh's forces to fight on the side of the British during the War of 1812.[29] McCoy was aware of the national defense argument as well. In *Remarks on the Practicability of Indian Reform,* he mentions that the ideal location for Indian removal was the area just west of Missouri because it was removed from direct contact with the Canadian and Mexican borders and on the periphery of the United States. In McCoy's mind, the location would render the Natives safer from the manipulations of British or Mexican agents.[30]

For those who favored removal, the most important element of the area immediately west of Arkansas Territory was the presence of an undesirable region even farther west than that. John C. Calhoun used this argument in his *Plan for Removing the Several Indian Tribes West of the Mississippi.* In the plan, the secretary of war put words in the Cherokees' mouths when he claimed that they would be willing to immigrate to this area, "as they have evinced a strong disposition to prevent the settlement of the whites to the west of them."[31] He went on to state that "one of the greatest evils to which [the Indians] are subject is that incessant pressure of our population, which forces them from seat to seat, without allowing time for . . . moral and intellectual improvement. . . . To guard against this evil, so fatal to the race, there ought to be the strongest and most solemn assurance that the country given them should be theirs . . . without being disturbed by the encroachment of our citizens."[32] Ultimately, Calhoun and McCoy's arguments convinced Congress to pass the Indian Removal Act in 1830, thereby giving President Jackson permission to conduct treaties with the eastern tribes, purchasing their lands and relocating them west of the Mississippi River.

In 1836, the Committee on Indian Affairs took up the same argument on the undesirable boundary region of the newly designated Indian lands. The committee described the region of "woodless plains" to the west of the site for Indian removal in a manner reminiscent of James and Long. At this time, the western border of the United States was the Rocky Mountains, U.S. territory extending only as far as the headwaters of the rivers draining into the Mississippi River, but the committee's report named the Indian reserve's western boundary as "an open, almost woodless plain, four or five hundred miles in width, in which on account of the scarcity of wood, no human being ever had a permanent residence."[33] The absence of human settlement on the plains was the selling point for relocation. As the Committee on Indian Affairs stated, "With this uninhabitable region immediately to the west of the Indian Territory, [the Native tribes] cannot be surrounded by a white population," and more importantly to the Congressmen, relocation there would place the Indians on the nation's perimeter, "and in a place which will ever remain an outside."[34] The notion of the "Great American Desert" was key because it would preclude expansion of the United States to the west, leaving the proposed Indian Territory in a non-threatening location.

Though the U.S. government had supported voluntary removal for more than twenty years without widespread results, the government under President Andrew Jackson began aggressively working to relocate the eastern tribes west the same year the Indian Removal Act was passed, 1830. But Isaac McCoy's claim that removal was vital to "civilizing" the Indians rang hollow when applied to major southern tribes, which had already taken on many elements of white culture. The Cherokees had elected officials as well as a newspaper written in English and in the Cherokee syllabary that had been created by George Guess, now known as Sequoyah. Many of the Five Tribes had adopted Christianity, wore Euro-American dress, lived in mainstream dwellings, and sent their children to American schools. The reasons for removal were different in the South. As non-Indian farmers played out their soil, planting crop after crop of cotton, the soil lost its nutritional value and yields fell. Farmers clamored for new lands to exploit; removal would open up tribal domains to non-Indian agriculturalists who dreamed of large plantations producing "King Cotton."

Jackson's government agents concluded treaties with the Five Tribes of the South with much controversy, as the treaty signings often included spirituous liquors, the prohibition of missionaries from the proceedings, and at times the circumvention of tribal governments to obtain signatures. Some tribes were fractured irreparably, such as the Cherokees, who passed a law calling for the execution of anyone who placed his mark on a treaty dissolving tribal rights to land. A tribal council ordered the execution of the signees of the Treaty of New Echota, which the Cherokee legislature never ratified.[35] Likewise, the Muscogee people, commonly known as the Creek Indians, executed their own tribal chief for signing a removal treaty.[36] Some residents of every tribe simply refused to move to the designated areas and fled to the mountains, or in the case of the Seminoles, the swamps. Thus, there remain eastern branches of most of the Five Tribes while their western cousins have lived for decades in modern-day Oklahoma. Further, the removed tribes faced stiff resistance from the tribes already inhabiting the territory set aside for removal. A fierce war raged between the Osages and the Cherokees during the first decade after removal.

The 1830s through the 1840s were, for the most part, wet years in the southern plains. These lush decades would see a new interpretation of the region: the Great American Desert became a land of promise. This competing interpretation can be largely attributed to three individuals: a freighter, an author, and a military officer.

THREE

Between the Droughts

The first great drought of the nineteenth century to hit the southern plains tormented the Long Expedition but passed with little notice among non-Indians farther east. The Native people coped with it as previous nomads had: they moved. The following thirty years proved to be much more humid, and the buffalo migrated back to the southern prairies, allowing the Comanches and Kiowas to return to their large domain and continue the way of life they had enjoyed for more than one hundred years. The grasslands resumed their ecological niche as a flight path for migratory fowl; the playas glistened with water in the fall when the geese and ducks flew south through the spring as they returned to their northern nesting grounds. The Plains Indians again took up their trade systems with the Pueblos and the Wichitas, and the southern grasslands appeared to be anything but a Great American Desert. The climatic shifts that influenced life on the southern plains remained unnoticed by those east of the Mississippi River.

Regardless of the environmental changes on the plains, no one stopped to think that Long had traversed the area during an incredibly severe drought, the effects of the arid climate amplified by the expedition's poor provisions. As noted earlier, the expedition's report labeling the region a "Great American Desert" was the foundation for the maps of countless atlases and textbooks; its regional label remained the scientific title for the region until the late 1850s.[1] It would take another twenty-four years after the Long Expedition before a sickly traveler turned scientist named Josiah Gregg challenged the permanence of the expedition's appraisal.

Born to a Tennessee farm family in 1806, Josiah Gregg excelled at his schoolwork and could often be found estimating the height of trees with a self-made wooden quadrant while his brothers were fishing after a long week of work.[2] As an adult, Gregg tried his hand at teaching on the Missouri frontier, then at medicine and law, but his curious mind was not at home in the courtroom or the legal library, and his health deteriorated with bouts of consumption, today known as tuberculosis, and chronic dyspepsia, or recurring acid reflux. Eventually, Gregg could not muster the strength to leave his room.

Gregg's physicians advised him to travel to the high prairies farther west, where the dry air would help soothe the effects of his lung disease. Through the connections of his older brothers, Gregg was able to join a wagon train bound for Santa Fe in May 1831, just nine years after that lucrative market was opened for trade between the United States and Mexico. Amazingly, within two weeks of his departure from Missouri, his health improved enough that he was able to resume normal activities. Perhaps the intellectual stimulation of the curiosities of the trail, a diet of wild game and beans, or the more arid conditions revived the studious traveler. Gregg enjoyed the organization of the caravan with its election of a train boss, its movement along four adjacent paths to cut down on the amount of dust eaten by those at the back of the lines, and the efficiency of the teams. The camaraderie of the trail, the hunts, and the adrenaline rush of Indian raids greatly excited Gregg's adventurous spirit. Like many others since, Gregg fell in love with the large vistas, beautiful sunsets, and cool evenings of the plains.

As a participant in the traffic of the Santa Fe Trail, Josiah Gregg was certainly qualified to give a firsthand account of environmental conditions in the southern plains after the nineteenth century's first drought. He had, of course, heard of the dry aspects of the plains, if not from the atlases of the time, certainly from his brothers, as well as from those with whom he traveled. On Gregg's first trek over the trail, he mentioned the anxiety associated with leaving the Arkansas River on the Cimarron Cutoff, a fifty-mile crossing to the Cimarron River without intervening access to fresh water. He described that crossing as "the scene of such frequent suffering for want of water. It having been determined upon, however, to strike across this dreaded desert [in] the morning, the whole party was busy preparing for the 'water scrape' as the droughty drives were very appropriately called by prairie travelers."[3] The crossing occurred without the usual suffering, as a heavy summer rain provided so much water that one small creek was swollen into a torrent, blocking their path. Aside from the delay of retrieving merchandise swept down the watercourse from one capsized wagon, the freighters continued to the Cimarron with little difficulty.

To read *Commerce of the Prairies* is to be impressed with how well read and observant Gregg was. He mentions Baron Alexander Von Humboldt's error in designating the Pecos River east of Santa Fe as the source of the "Red River of Natchitoches."[4] A similar mistake plagued Long during his expedition. Mistaking a southward-flowing watercourse east of Santa Fe for a tributary of the Red River, he instead descended the Canadian and hence failed to locate the source of the border river.

Gregg's depiction of the plains contradicts the Long team's descriptions so starkly that it is hard to imagine that they are describing the same region. In 1839, Gregg took another trip to Santa Fe, this time embarking from Van Buren, Arkansas, and following the Canadian River into New Mexico. Whereas the Long report placed the eastern border of its desert at Webbers Falls, Gregg depicts the same area as a "landscape beautifully variegated with stripes and fringes of timber" with small herds of buffalo scattered in between, a picture "truly delightful to contemplate." He called the area Spring Valley. In the same paragraph, he claimed the valley contained "numerous spring-fed rills and gurgling rivulets that greeted the sight in every direction, in whose limpid pools swarms of trout and perch were carelessly playing. Much of the country indeed over which we had passed was somewhat of a similar character."[5] In complete honesty, Gregg later claimed the high plains were "too dry and sandy" to be fit for cultivation, but he asserted that the desert region was much farther west than Long located it. It is helpful to remember that Gregg crossed the region four times, all during years of average to abundant rainfall.

The amateur geographer continued to use terms like "desert" for the high plains, but he was quite optimistic about the area to the east. Gregg contradicted Long's claim that the area from the ninety-sixth meridian to the Rocky Mountains was a "wide sandy desert," pointing out that the landscape from the Arkansas border to the Cross Timbers was "interspersed with prairies and glades, many of which are fertile," and writing that the source of the Colorado River in west Texas was "delightfully watered" but bordered the "immense desert region of the Llano Estacado."[6] Gregg's greatest contribution to the Great American Desert concept was to move it west, criticizing the federal government for granting its eastern portion to the removed tribes: "We have here, then, along the whole border [of Arkansas and Missouri] a strip of country, averaging at least 200 miles wide by five hundred long . . . affording territory for two states respectable in size, and though more scant in timber, yet more fertile, in general, than the two coterminous states of Missouri and Arkansas. But most of this delightful region has been ceded to the different tribes of the Frontier Indians."[7]

Gregg began writing an account of his travels on the Santa Fe Trail late in 1843, using his diary entries and letters to jog his memory. After rejections from several publishers and extensive editing, Gregg's book finally made print in July 1844.[8] *Commerce of the Prairies* represented a major shift in tone from the previous literature, portraying the nation's interior as more suitable to settlement.

In many ways, Gregg was ahead of his time. His optimistic report on the geography of the southern plains fed into the growing popularity of westward expansion so commonly described and excused as Manifest Destiny. Ten years after publication of *Commerce of the Prairies,* Congress opened portions of the Great American Desert by passing the Kansas-Nebraska Act, requiring the Frontier Indians of those territories to relocate once again, this time southward to Indian Territory. Furthermore, Gregg had identified the possibility of climatic shift on the southern plains. He noted that "the High Plains seem too dry and lifeless to produce timber; yet might not the vicissitudes of nature operate a change likewise upon the seasons? Why may we not suppose that the genial influences of civilization—the extensive cultivation of the earth—might contribute to the multiplication of showers, as it certainly does of fountains?"[9] This concept would become highly influential forty years later as another major misinterpretation of the region took hold in the popular imagination, this time characterizing the plains as a garden growing in the desert.

Washington Irving promoted the southern grasslands in much the same way as Gregg. Like Gregg, Irving became so sickly as a young man that his siblings urged him to leave his home to seek a healthier climate. The family sent him to Europe with the hope that his health would recover. He boarded a ship bound for France in 1804, and such was his condition that the ship's captain remarked to another passenger, "There's a chap who will go overboard before we get across."[10] In Europe, the young Irving's health did recover, and he enjoyed a year-and-a-half tour of France, Italy, and London before heading back to New York City in 1806. After a short stint as a lawyer, which included participating in the defense of Aaron Burr in his trial for treason, Irving turned to writing. In 1809 he published a spoof titled "A History of New York," which poked fun at the upstate Dutch and their economic activities. This wryly humorous piece won him acclaim in the United States and in Europe and established the term "Knickerbocker" as a moniker for a New Yorker.

After the War of 1812, Irving once again headed for Europe, this time to attend to family business interests in London. He traveled to Liverpool on one of Commodore Stephen Decatur's ships that were ultimately bound for the Mediterranean to tackle the Barbary Coast pirates for the second time, and he remained in England as the U.S. Navy extracted treaties from the North African principalities. While in the British Isles, Irving realized that the family business was unsustainable, so he took to writing in earnest and produced his most famous short story, "Rip Van Winkle," in 1819. While in London, Irving hobnobbed with famous English gentlemen such as Sir Walter Scott and Benjamin Disraeli.[11] For seventeen years Washington lived in Europe on the proceeds from his publications.

While Stephen Long was suffering through his exploratory journey through the southern plains in 1820, Irving traveled to Paris, where he remained for a year before heading on to Italy, and ultimately, Spain. While in Madrid, he researched and wrote *The Life and Voyages of Columbus*. With wanderlust still tugging rather vigorously at Irving's sleeve, he wended his way through Cordova to Granada, Malaga, Gibraltar, and Cadiz, gathering information for another book, *A Chronicle of the Conquest of Granada*. The American spent the summer of 1829 enjoying the hospitality of the governor of Alhambra, who had invited Irving to live in the grand palace. He passed the sunny days and cool nights adding the finishing touches to his two manuscripts and strolling through the promenades. The Fountain of Lions in the street below inspired a later manuscript, *The Alhambra*.[12] At the end of the summer, he received news of his appointment as secretary to the United States's diplomatic legation in London. He assumed his post in September of 1829; then, in the last months of his service, he worked as the Chargé d'Affaires for the newly appointed legate, Martin Van Buren.[13] After seventeen years, Irving finally felt a longing to see his homeland, and he made up his mind to quit the diplomatic corps and return to the States.

Irving departed from Le Havre, France, in April 1832 to return to the land of his birth. On the passage, he met and struck up an immediate friendship with Charles Latrobe, a fellow traveller of Europe who had recently been

hired as tutor and traveling companion to the nineteen-year-old Count
de Pourtalès. The two were bound for America on a "tour of curiosity and
information."[14] As the ship arrived in New York City, Irving offered to host
the young count and Mr. Latrobe.

Irving returned to a United States marked by political strife and expansion: Jackson was embroiled in a political battle over renewing the national
bank, South Carolina was discussing nullifying the federal taxes on imports
and exports, and Indian removal was proceeding, as Congress had passed
the Indian Removal Act in 1830. The now middle-aged Irving had a strong
desire to travel across his native country, and the count and Latrobe opted
to join him. The three journeyed across the northeast, and while on board
a lake steamer heading from Buffalo, New York, to Detroit, by chance met
Henry L. Ellsworth. Earlier that summer, Secretary of State Lewis Cass had
offered Ellsworth an appointment as Indian commissioner and instructed
him to organize an expedition to "examine the country set apart for the
emigrating Indians."[15] Ellsworth asked Irving to join his expedition, and
the recently returned expatriate quickly assented, excited about the opportunity to experience the West firsthand. Latrobe and the Count agreed to
follow along at their own expense.

On its way from Lake Erie to Fort Gibson, the group traveled to Saint
Louis in September, where Irving's curiosity led him to visit the jail cell
of recently captured Black Hawk, the leader of the Sauk and Fox Indians
who had fought a losing struggle from May to June of 1832 to retain their
homeland against the United States. The renowned author described the
chief as an "old man upwards of seventy with aquiline nose—finely formed
head—organs of benevolence." Irving's terse lines hint at melancholy as he
describes Black Hawk, his son, and brother-in-law suffering wrist and ankle
chains in the cell.[16] From there the expedition traveled southwest to Fort
Gibson, where its tour of the prairie began. Irving, Latrobe, and Ellsworth
all published accounts of their journey from the western military post along
the Cimarron River to near modern day Oklahoma City, and returning to
the fort via the Canadian River. Irving's account, *A Tour on the Prairies,*
was the most publicized and widely read.[17]

In *A Tour on The Prairies,* Irving characterizes the tallgrass prairies of
central Indian Territory in glowing terms. Of course, the above-average
rainfall that had inundated the region during the year of his journey, 1832,
influenced his perception. Just as Josiah Gregg had experienced on the Cimarron Cutoff, the heavy rains during Irving's tour impeded parts of the

journey. Irving describes the impact of heavy thunderstorms on the creeks: "The brook which flowed peaceably on our arrival, swelled into a turbid and boiling torrent, and the forest became little better than a mere swamp."[18] The party's horses trudged through grasslands "rendered spongy and miry by the recent rain."[19] In all, the expedition experienced four rainstorms in the span of twenty-eight days.

Irving uses a lot of ink in reciting the successes of the expedition's hunters in bagging game. Their meals rarely lacked meat during the journey, quite a different experience than the Long Expedition endured. In the words of Irving, "The surrounding country, in fact, abounded with game, so that the camp was overstocked with provisions."[20] The landscape's munificence eventually became a liability, as the less experienced men began to waste their supplies. The military personnel who traveled with the commission were rangers, volunteers who signed up for a one-year hitch. Many of these rangers left whole sides of meat at their deserted camps, thinking they could easily replace it by hunting along the way.[21] Other members of the expedition also used the abundance of game as a pretense for consuming their provisions quickly with little regard for rationing. According to Ellsworth, Irving was as bad as the rangers when it came to consuming the provisions.[22] Late in the trek, the hunters' luck abandoned them, and the party suffered from want.[23] The excesses and waste involved in the party's hunts triggered an emotional response from Irving, at times echoing the environmentalism of the Transcendentalists back east. When the expedition's Métis cook killed a polecat to embellish the camp meal, Irving referred to the event as the "Foul Murder of a Skunk." He discusses with contempt the killing of an owl purely as a joke and the unfortunate killing of three prairie dogs for sport, and he offers remorse for participating in the killing of a buffalo bull: "A rifle ball, however, more fatally lodged, sent a tremor through his frame. He turned and attempted to wade across the stream, but after tottering a few paces, slowly fell upon his side and expired. It was the fall of a hero, and we felt somewhat ashamed of the butchery that had affected it."[24]

Irving's prairie is nothing like Long's desert. Although he uses the term "waste" on two occasions to describe the topography, he never applies the word "desert." The reader is left with a vision of the newly created Indian Territory as a land of plenty. In fact, Irving uses biblical imagery to illustrate the desirability of the region: "It seems to me as if these beautiful regions answer literally to the description of the land of promise, 'a land flowing with milk and honey'; for the rich pasturage of the prairies is calculated to sustain herds of cattle as countless as the sands upon the seashore."[25] The region Irving describes is the same area, two hundred miles from the western borders of Arkansas and Missouri, that the Long group considered uninhabitable. The commission did not travel onto the high plains and thus passed no pronouncement on the condition of its terrain, but another expedition would have the opportunity to evaluate the plains during an average year of precipitation.

In Mount Holly, New Jersey, just two months after Long and Bell had left Fort Smith at the end of their exploratory journey, a War of 1812 veteran named John Abert celebrated the birth of a son he named James. The infant would mature to become the leader of the next major military expedition across the Canadian River.

Feeling the pressure of his father and grandfather's valorous military service, James matriculated at West Point. The young Abert did not do well in his military courses but definitely excelled at drawing. Upon his graduation, he ranked fifty-fifth in a class of fifty-six and was placed in the infantry, which he endured for only a year before his father pulled some strings and obtained his son's transfer to the Corps of Topographical Engineers. Here his skill in sketching was put to good use diagramming bridges or portraying the countryside explored in the unit's expeditions.[26] In Abert's first exploratory excursion, he helped survey the southern shores of the Great Lakes in 1843–44.

In 1845 Lieutenant James W. Abert joined John C. Frémont's reconnaissance expedition, which traveled to Bent's Fort in modern-day southeastern Colorado with orders to reconnoiter the Canadian River in the heartland of the Comanche and Kiowa tribes. At Bent's Fort, Frémont decided to take the bulk of his unit to California, as relations between the United States and Mexico worsened and war loomed on the horizon. Frémont gave command of the Canadian River expedition to Lieutenant Abert, who led his party through the same stretch of the Canadian River that the Long Expedition had travelled. But Long's party had journeyed through the southern plains during a drought, and 1845 was a year of average moisture. Not surprisingly, Abert's commentary on the region is drastically different.

Abert's group bade farewell to the dashing Frémont on August 13 near Bent's Fort. His party consisted of thirty-five men, wagons, and eight "prairie fed beeves," just in case the expedition failed to encounter enough game to keep the men fed.[27] Abert's writings reveal that he was more attuned to the cycles of nature than was his predecessor, Long. The lieutenant described the dryness of Timpas Creek in present southeastern Colorado and qualified his statement by mentioning that "the drought this year [in the environs of Bent's Fort] has been great."[28] The next day he related the effects of the arid conditions on the team's supplies, describing how a wooden wheel simply fell off the axle of one of the wagons and attributing it to "the dryness of the atmosphere having caused the wood to shrink."[29] The party traveled across Raton Pass, hitting the Canadian River shortly after descending into present New Mexico. Although they experienced a few rainstorms and showers, Abert referred to the region of eastern New Mexico and the Texas Panhandle as "a desolate desert," or "desert-like," and wondered at the Native people's ability to subsist in such a barren environment.[30]

As the lieutenant traveled east, he noticed a change in the topography and climate, once referencing Josiah Gregg's *Commerce of the Prairies* and evoking the author's name elsewhere in his report.[31] Like Gregg, Abert placed the "Great American Desert" much further west than Long did. As Abert's group departed from the Caprock Escarpment in the vicinity of today's Texas Panhandle, Abert began to notice a change in the landscape and marked the occasion with an entry in his journal: "it was evident that we had passed the great desert."[32] The Caprock Escarpment is approximately 450 miles west of Webbers Falls, where Long claimed the desert began.

Abert's party found plenty of game: buffalo, deer, turkey, even water fowl, and resorted to killing only two of their eight cows on the journey. They rarely worried for fresh water and often described the terrain in glowing terms. In present Blaine County, Oklahoma, Abert reported "a small creek, which in several places along its course, widened into small lakes of five feet in depth," and a day later he noted "plenteous rivers and wide skirted meads."[33]

Like Gregg and Irving before him, Abert reached a conclusion about the possible settlement of the region that contrasts acutely with James's. Abert described "a country so beautiful, abounding as it does with timber, with water, with, in fact, all the allurements which would induce man to frequent it."[34] History could have been much different had the Long Expedition traveled through the southern plains during a wet year, but, of course, this was not the case. It took decades to convince most Americans that the interior grasslands were not a desert, but in the meantime, the young nation's energies were directed at expanding across the continent, and this too may have influenced its perception of the plains.

By the spring of 1848, Gregg was living his dream of collecting plant specimens and experiencing new lands and new people after joining a botanical expedition through Mexico. The Treaty of Guadalupe Hidalgo, signed on February 2, 1848, had ended the war, making travel safer for the herbologists, and by June 30, 1849, the group was on the shores of the Pacific Ocean at Mazatlán. From there Gregg sailed to San Francisco to be reunited with Jesse Sutton, his old partner from his Santa Fe Trail days. Here his lust for adventure again overcame him, and Gregg headed east for the gold mines of the Sierra Nevada. Now forty-three years old, the intrepid journeyman agreed to head a seven-man team across the Coastal Ranges in search of the ideal bay to create a more direct route from which to transport supplies to the gold mines of the Sierra Nevada. The current route from San Francisco was long, and a port further north could cut days off the trip. This expedition appealed to Gregg on many levels: the adventure of trailblazing, the chance to see new plants—including the giant redwoods of California—and the opportunity to establish a new business selling expensive goods to the gold miners. In fair weather, the trip should have taken eight days, but Gregg's insistence that he gather a sample of every unique plant he came across slowed the group's progress. It took a full month to cross the coastal mountains, and as the party's supplies ran low, the men began to bicker. They eventually found a bay sufficient for their plans and the group headed south toward San Francisco, but the difficulties of little food, approaching winter, and strained relationships encouraged the team to split up. Gregg's group headed inland toward Clear Lake, where after suffering a severe injury from a bad fall from his horse, the self-taught scientist and author perished and was buried on the spot. After the others completed their journey and returned to the scene of Gregg's demise, they could not locate the grave.[35] Perhaps such a wilderness burial is appropriate for one who loved the exhilaration of travel and held a true passion for plant and animal life.

Gregg, Irving, and Abert were some of the first to see the grasslands in a new way, not as a "Great American Desert" but as a region of promise. Gregg's work as a businessman appealed to those looking for economic opportunity in the West. Irving, as much a European as an American, saw the region with fresh eyes, and his popularity ensured that this new description of the region would be widely read in Europe as well as the United States. Abert's report did not have the same circulation among the reading public as did the accounts of the other two writers, but it was available to those in government offices and, as a military report, carried the weight of a government stamp.

In the meantime, a combination of philosophical, political, and economic forces coalesced to support the settlement of the plains. The burgeoning concept of Manifest Destiny influenced individuals such as Thomas Hart Benton and William Gilpin to promote the settlement of the grasslands. Benton, a U.S. senator from Missouri, was a strong supporter of transcontinental railroad construction, and came to laud the possibilities for settlement of the plains. He spoke of the region formerly known as the Great American Desert as a "sylvan paradise" of sufficient humidity to grant "fertility to this region." With the aid of his son-in-law, John C. Frémont, Benton sought to reverse the bad publicity the region had received.[36] William Gilpin, who had traveled with Frémont to Oregon in 1843 and had studied under the highly respected Prussian naturalist Alexander von Humboldt, also scorned the negative publicity of the American steppe. Gilpin, in *Mission of the North American People,* plainly stated that "the PLAINS are not deserts," but instead the most advantaged area between the Appalachians and the Rockies.[37] Benton and Gilpin, both writing in the 1850s, were also joined by interests that formed during another divisive crisis of that same decade.

In opposition to the spread of slavery allowed by the Kansas-Nebraska Act, many promoted white migration to the centrally located plains. The foremost agent in recruiting settlers for Kansas was the New England Emigrant Aid Company, a booster that hired writers to describe the grasslands in glowing terms and convinced many that they could do God's will and reap a healthy reward for their efforts, if only they would move west onto the treeless prairies.[38] To make its invitation more compelling, boosters had to dispel the myth of the Great American Desert. The first publication to use this strategy was E. E. Hales's *Kanzas and Nebraska,* out for circulation in 1855. Hale claimed that the area west of the Missouri border may have been a desert, but that these conditions would lessen as civilization expanded into the region.[39] That same year, two employees of the American Reform Tract and Book Society, Charles Boynton and T. B. Mason, released *A Journey Through Kansas with Sketches of Nebraska* in which they claimed the desert was a complete myth.[40]

Republican party interests were similarly motivated but slightly more complicated than those of the boosters. The Republican platform included the building of a transcontinental railroad, a Homestead Act, and opposition to the spread of slavery into the territories. The popularization of the plains, and the economic growth that would come with it, benefitted all these political initiatives. For so many reasons, there were a growing number of entities who desired opening the plains for settlement, and all described the region in favorable terms. But the momentum of these proponents of western expansion was soon to be disrupted by the most severe drought to hit the southern plains in the previous 250 years.

FOUR

Drought and Buffalo Migration

With adequate moisture and a bountiful harvest, the year 1853 appeared to promise peace to Indian agents of the southern plains.[1] The Osage Indians journeyed close to eight hundred miles out onto the grasslands to hunt buffalo, uncharacteristically, without a single conflict with the Plains tribes.[2] The Cherokees, beleaguered by the Plains nomads, sought out peace with the tribes of the High Plains, and the U.S. government took its own active role in promoting peace with the Kiowas and Comanches by negotiating the Treaty of Fort Atkinson.[3] Even the Comanches and their hated enemies, the Texans, seemed to be moving to a more tranquil relationship. That same year, Texas Indian agent Robert Neighbors traveled to the plains to visit the Southern Comanche bands, and noted in a report to the Commissioner of Indian Affairs that "during my stay with them, I could discover nothing of any hostile or warlike disposition; and I have every reason to believe that we shall have a season of uninterrupted peace on our frontier."[4]

The very next year, amicable relations deteriorated between the High Plains Kiowas and Comanches and the relocated tribes of Indian Territory and their ally the United States. In the spring of 1854, Indian agents reported that the Kiowas, Comanches, Plains Apaches, Arapahoes, and Cheyennes had gathered at the Pawnee Fork of the Arkansas River in present-day Kansas to form a war party with the intent of wiping out "all frontier Indians they could find on the plains," including the Native peoples of eastern

Kansas and Indian Territory, who made forays onto the plains to hunt buffalo.[5] Two years later, Comanche raids upon reservation Indians, Hispanic Tejanos, and Texans increased dramatically. During the following five years, relations between the Texans and the Plains Indians continued to worsen. Granted, the relationships between Plains Indians and their eastern neighbors were already tenuous at best, but the question remains: why the complete turnaround in what seemed to be improving relations?

Scholarly theories on the cause of increased frontier hostilities during the 1850s range from blaming the removal of troops from their frontier posts on the eve of the Civil War, to competition for buffalo hides, and simply to increased traffic through the southern plains tribal domains.[6] It can certainly be argued that the buffalo migrations were more important to tribal warfare than previously thought, since buffalo movements directly affected the diplomatic choices of the southern plains tribes. Further, this period of excessive drought exaggerated the effects of the overland trails on the ecosystem, pushing the bison herds to the east and encouraging the Plains tribes either to accept reservation life and commit to reliance on the federal government or to increase their raids on neighboring people and incur warfare with the United States.

Of course, drought had affected the migrations of bison herds for thousands of years.[7] As the lack of rainfall withered the grasses on the high plains, the herds either moved eastward to locate sufficient pasture to sustain their population, or starved.[8] People residing on the interior grasslands who were dependent on bison hunting had to rely on trade and raiding to supplement their diet with agricultural products. During average conditions, bison herds provided plenty of hides to trade with agricultural people like the Wichita for corn, but when the herds moved far enough away, the southern plains foragers were faced with few peaceful options. During times of mild drought, the Plains people could migrate to the edges of the high plains and camp at one of the many springs that bubbled up at the base of the Caprock Escarpment, or travel even farther east, relying on hand-dug wells for water during these eras of sparse rainfall.[9] Paleo-foragers utilized farming and hunting to subsist on the plains, but what would a culture specializing in hunting do during times of severe drought? They could migrate as far east as the herds wandered and risk infringing on the hunting grounds of other tribes, they could raid the agricultural villages in their vicinity for necessities and bartering material with which to obtain corn and other staples, or a little of both.

George Catlin, *Buffalo Hunt, A Numerous Group*, 1844. Day and Haghe, hand-colored lithograph. Courtesy of the Buffalo Bill Center of the West, Cody, Wyoming. Gift of Mrs. Sidney T. Miller, 21.74.7.

The U.S. government's removal of the eastern tribes to the Trans-Mississippi West blocked the Plains Indians' movement to the east during times of drought. As Comanche and Kiowa hunters moved east with the buffalo, they came into conflict with the government-supported tribes of Indian Territory, but, unable to obtain buffalo hides, the Plains Indians had little to offer their neighbors in trade. Thus, they resorted to raiding, which, of course, elicited a response from the federal government. To understand just how damaging the combination of a severe drought and the presence of federally protected Indian tribes on the eastern prairies was to the Kiowas and Comanches, it is necessary to look at the prehistoric relations of high plains Natives with their neighbors, investigate the circumstances that propelled the Comanches and Kiowas to leave their homelands farther north to journey southward onto the southern plains, and explore the severity of the 1850s drought and its effects on the animal and human populations of the region.

Precontact Plains hunter-gatherers were dependent on trade with their eastern and western neighbors. As Pueblo populations grew, they tended to overhunt their immediate vicinity and were forced to become more specialized in growing crops. The intensive hunting in the Pueblo region forced larger game like bison and elk to flee the region or be killed off, thus leaving the sedentary groups the option of either venturing onto the plains for a few big hunts or relying on trade with Plains nomads for meat. Under such circumstances the nomadic people of the grasslands came to rely more and more on hunting buffalo, for they could fall back on trade with the Pueblos to the west or the Caddos to the east to procure the much-needed carbohydrates of maize, beans, and squash.[10]

Hostilities between the agricultural tribes and the Plains foragers were probably rare because there was much more for both groups to gain from peaceful trade than from belligerence. Prior to Spain's reintroduction of the horse to North America, the Plains tribes were numerically and militarily inferior to the Pueblos and Caddos, with their easily defensible dwellings, large villages, and confederated allies. If the nomads ventured an unmounted attack on the agricultural villages, they would risk losing their trade partners for good, while paying a heavier human price than their defending counterparts. When a member of a sedentary tribe was wounded, his people could nurse him back to health while another member of the tribe took up his work in the fields. But when a nomad was wounded, he had to be able to keep up with the tribe or he would endanger the well-being of his people. Raiding was equally risky without mounts due to the difficulty of a quick escape with stolen goods in hand. The agriculturalists would have plenty of time to organize a war party and track the slow-moving raiders down.[11]

During the 1660s, when the Spanish prohibited Pueblo trade with the Apaches, the only alternative for the Plains tribes to obtain Puebloan goods was to increase raiding. No doubt, the Pueblos found these conditions intolerable, especially as drought withered their fields. When the resulting Pueblo Rebellion of 1680 released the Spanish horse herds of northern New Mexico, the animals quickly spread throughout the plains.

By 1680, some eighty years after the arrival of conquistador Juan de Oñate and his entourage, the horses in the Spanish herds numbered in the thousands, yet only the neighboring Jumanos had incorporated the animal into their semisedentary culture.[12] The agricultural Natives had little need of horses and allowed them to become feral. Within one year of the revolt, the Mendoza-Lopez expedition made one of the first European observations of mounted Plains Indians. The party encountered Native horsemen near the Pecos River and reported the loss of a few horses to Indian stealth.[13] The eastern tribes of present Texas had horses by 1700, and the equines had diffused as far north as the eastern slopes of the Canadian Rockies by 1750.

Once the Plains foragers procured mounts, they experienced a technological revolution. They could more easily locate and hunt buffalo herds. On mounts, hunting the burly animals, though still dangerous, was just a matter of riding into the galloping herd and dispatching a well-aimed arrow or spear behind the shaggy creature's shoulder blade. With horses, the nomadic tribesmen were also able to move more quickly, allowing them to keep up with the movements of buffalo herds across the grasslands. The easier access to bison allowed Plains hunters to trade for more goods, if not with the Puebloans, certainly with the Caddos or with non-Indian traders. Increased trade, in turn, allowed the populations of the Plains tribes to grow. This population growth, along with the tribes' newfound maneuverability, gave them the advantage over their agricultural neighbors. Now a raiding party could speedily escape with the fruits of its attack, making off to the high plains, where its knowledge of the limited water sources was the difference between life or death.[14]

The Comanches entered the plains at just the right time to take advantage of this mobile way of life. In the sixteenth century, the group that would become the Comanche tribe was still part of the Shoshone people, who lived in the Great Basin of present northern Utah and the adjacent sections of its neighboring states. At some point, a few bands crossed the Rockies, leaving their ancestral domain along the Snake River for the eastern slopes of the Rocky Mountains in modern Colorado and Wyoming.[15] Within three generations, they moved south. Perhaps this migration was in search of horses and game, or perhaps they felt the pressure of the Sioux groups, now armed with European firearms, who were pushing onto the northern plains from the western Great Lakes region. At any rate, by 1706, the term *Los Komantcia* began showing up in Spanish documents and by 1735, with the easing of trade restrictions, the Comanches regularly exchanged goods with the Spanish and the Pueblos of the upper Rio Grande.[16]

But relations between the Pueblos and Comanches soured in the late 1740s. The New Mexican governor reinstated the prohibition on trade between sedentary and nomadic peoples in 1746 and, not surprisingly, the Comanches attacked the Pecos Pueblo, initiating a flurry of raids on the Pueblos within the following years.[17] Around the same time, some Comanche bands secured a peace agreement with their eastern neighbors, the Wichitas, and established trade with the French to procure the goods they had been obtaining from the Pueblos.[18] Also during the eighteenth century, the Comanches carried on an aggressive war with the Plains Apaches, pushing them west into the mountains of modern day New Mexico. By 1752 the Comanches had secured the southern plains as their hunting grounds, which they shared with their new allies, the Kiowas and the Kiowa Apaches.[19]

The Comanches' reputed faithlessness in honoring treaties has been misunderstood due to their loose political structure. There was no supreme chief of the extended tribe; a leader of one band could not vouch for any other group of Comanches. Thus, a truce concluded with one band of Comanches would not preclude an attack by another.[20] All through the seventeenth and eighteenth centuries, some groups of Comanches truced with their neighbors while other Comanche bands raided them. To further complicate matters, members of one band could simply pack up and move in with a warlike band if they did not agree with a treaty negotiated by his own.

As a nomadic Plains people, the Comanches depended on hunting, trading, and raiding for their livelihood. To a large extent, their success in the former two initiatives influenced their engagement in the latter. During times of plenty, the Comanches had meat for their families and hides to trade for other necessities and desired objects. During periods when game was scarce, there were few hides to trade for these products. At such times, the Plains warriors relied on raiding their neighbors for livestock: cattle for meat and horses and mules for bartering in the acquisition of trade goods. To engage in this lifestyle, they needed a large domain.

During the 1800s the *Comancheria* was extensive, stretching from the Arkansas River in the north to the Balcones Escarpment of Texas in the south, and from the Cross Timbers of present-day central Oklahoma in the east to the Pecos River in the west.[21] There was good reason for protecting such a vast hunting range. The southern plains bison herds migrated east during the spring and summer to take advantage of the tallgrass prairies, which provided succulent nutrients only during those seasons. When the tall grasses turned brown, the ungulates moved back west during the approach of colder weather to take advantage of the high plains short grasses, which continued to grow throughout the winter.[22]

On a more local level, this broader migration coincided with seasonal movement from river basins to uplands. During the colder months, the bison moved closer to the river and creek valleys to take advantage of the sheltering stands of trees in order to escape the effects of the cold winter winds. In the spring, when the tall grasses lining these valleys began to sprout, the bison were ready to take advantage of them. As the warmer months came, the buffalo moved up onto the intervening highlands to graze on the short grasses.[23] In a region so vast and unbroken by natural barriers, the bison could usually find green pastures somewhere, and with horses, the Comanches could usually find the bison.

The welfare of the Comanche people depended on the buffalo hunts for more than meat. A burgeoning trade in buffalo hides was spreading to both their east and west. In the 1700s, many tribes attended trade fairs to barter agricultural goods for others' buffalo hides, or to trade livestock with industrial goods from the Spanish.[24] These trade centers proved so lucrative for the Spaniards that in 1786, Juan Bautista de Anza began a policy that allowed New Mexican traders, who came to be known as *Comancheros,* to journey out onto the plains to trade with the Comanches. De Anza was motivated not only by profit, but also by the assurance of peaceful relations with the Plains tribe, assuming that a lack of trade would result in an increase of raids on New Mexican communities.[25]

The Comanches became the primary supplier of horses and buffalo hides through a fluid system of treaties, truces, and mutual understandings with the Spanish colonists. While they maintained peace with the New Mexicans, they obtained slaves and horses from Mexico and Texas. If relations fell apart with the New Mexicans, the Plains nomads shifted their trade elsewhere and commenced raiding in New Mexico for captives and livestock. Occasionally, they raided one ranch for horses and then traded these animals to another settler some miles away.[26]

American traders also established posts on the plains to take advantage of the huge Indian supply of buffalo robes, which were in demand in eastern markets, and further integrated the Comanches into a market economy.[27] To some degree, they became dependent on manufactured goods as, similarly, they had always been dependent on their neighbor's grain. But the return of drought to the plains triggered a chain reaction that disrupted the fragile balance the residents of the grasslands had established during the previous forty years. Drought began depriving the ecosystem of water in the late 1840s, but the severely arid conditions hit with full force in 1854 and continued for the next ten years.

In the mid-1800s, the problems for the Plains tribes continued to mount. The results of the Treaty of Guadalupe Hidalgo in 1848 were not in their favor. In the treaty, Mexico transferred to the United States all of the southern plains and the territory westward to the Pacific Ocean, ending the ability of the Comanches and other Plains peoples to play the Mexican and U.S. governments against each other, seeking better trade agreements or maintaining good relations with one while conducting raids against the other. In addition, traffic along the major trails crossing the southern plains increased as a result of the discovery of gold in California. One migration route for gold-seekers followed the Santa Fe Trail through the heart of Comanche land, and another followed the Canadian River. As many as 3,000 gold rush participants traveled along the trail in 1849 alone.[28] With the emigrants came diseases that devastated the Plains tribes. A cholera epidemic took the lives of more than half of the Comanche people and continued to sweep through the Kiowa, Kiowa Apache, and Southern Cheyenne during the first year of the gold rush migration.[29] These effects shattered the balance of Native lifeways on the southern plains.

Understandably, the river valleys were crucial to the welfare of the buffalo and Native populations. During the winter, humans, like the plains animals, used the river valleys' protective canyon walls or sheltering trees to escape the brunt of winter storms, to find forage, and to access reliable supplies of water. The growing Native population with its huge horse herds taxed the ecological stability of these valleys. Captain R. B. Marcy, during his military excursion to the region in 1852, described the effects of Comanche and Kiowa use of the North Fork of the Red River valley:

Vestiges of their camps were everywhere observed along the course of the valley, from the Wichita mountains to the sources; and the numerous remains of stumps of trees, which had been cut down by them at various periods indicated that this had been a favorable resort for them during many years. . . . From the great extent of surface upon which the grass was cropped at some of the camping places, and from the multitude of tracks still remaining, we inferred that they were supplied with immense numbers of animals; and they are undoubtedly attracted here by the superior quality of the grass and the great abundance of cotton-wood which is found along the borders of the streams, upon the bark of which they fatten their favorite horses in the winter season.[30]

During the spring and fall, buffalo and Natives evacuated the river valleys for the short grasses on the uplands, thus allowing the valleys to recover from the intense grazing by buffalo and horse herds. During the gold rush, non-Indian travelers began to use the river valleys to graze their livestock and provide firewood for their camps during the spring, summer, and fall. The continuous use of the grasses and trees by animals and humans left no opportunity for the valley flora to regenerate.[31] This lack of regeneration could certainly have influenced the bison herds' relocation to other valley systems where their demands were sustained.

Yet the Plains Indians had been complaining about the dwindling buffalo populations for some time. A frontier settler of Texas, Noah Smithwick, recalled a conversation he had with a principal man of the Penateka Comanche named Muguara during his stay with the tribe in 1839 and 1840. The Texan reconstructs an eloquent discourse by Muguara on the effects of non-Indian settlers' presence on his people's hunting grounds: "We have set up our lodges in these groves and swung our children from these boughs from time immemorial. When game beats away from us we pull down our lodges and move away leaving no trace to frighten it, and in a little while, it comes back. But the white man comes and cuts down the trees, building houses and fences, and the buffalos get frightened and leave and never come back, and the Indians are left to starve, or, if we follow the game, we trespass on the hunting grounds of the other tribes and war ensues."[32] By the late 1830s the Comanches were already blaming whites for the diminishing bison herds. In addition, by 1845 Kiowa and Comanche leaders were complaining to officers at Fort Atkinson that the New Mexico buffalo hunters were killing off too many buffalo.[33]

Non-Indian residents of the area were aware of the declining and migrating bison populations as well. An unknown author writing from Fort Gibson claimed that in 1822, buffalo abounded near Fort Smith, but that ten years later they could not be found within one hundred miles of the post.[34] The English adventurer Charles Latrobe, who traveled with Washington Irving from New Orleans across the prairies to Mexico, echoed this assessment, claiming that by the fall of 1832, he had traveled more than one hundred miles from "the remotest limit" of Arkansas Territory and had still not encountered a herd of bison.[35] As late as 1841, the herds still inhabited the far western tallgrass prairies. Concerned Indian agent William Armstrong substantiated the reports of buffalo presence on the high plains in the 1840s when he stated that since the eastern tribes had arrived at their region in the eastern prairies, the Osages were having to range farther west to locate the buffalo, and that this predicament would likely lead to hostilities between the Osages and the Plains Indians.[36]

Yet by the mid-1840s, the herds had moved eastward. George Ruxton, a twenty-five-year-old adventurer fascinated by the western frontier of the United States, journeyed through old and New Mexico with only a guide and pack animals. He crossed the high plains along the Arkansas River in 1846 and noted, "It is a singular fact that within the last two years the prairies, extending from the mountains to a hundred miles or so down the Arkansas have been entirely abandoned by the Buffalo."[37] Yet further east near Coon Creek, close to the one hundredth meridian, he saw buffalo so thick that hardly a patch of grass ten yards square was exposed to the sun.[38] In 1852, Captain Marcy, on yet another expedition to find the source of the Red River, also noticed the disappearance of the herds. He blamed the reduction of the large ungulate population on white hunters who indiscriminately slaughtered the animals for hides and for sport. Marcy claimed the buffalo herds had moved north of the Red River and were confined to a "narrow belt of country between the outer settlements and the base of the Rocky Mountains."[39] In fact, the presence of a single buffalo near the Cross Timbers in the vicinity of Fort Worth during that same year was so newsworthy that it was reported in the Clarksville, Texas, *Standard* newspaper.[40]

One might ask, why would the bison herds prove so difficult to locate? There was a series of factors that might have caused such wide dispersals. First, the buffalo hide trade had expanded in 1846 to an estimated 100,000 skins yearly.[41] The number of Native hunters who desired the meat and summer hides of two- to five-year-old buffalo cows greatly inhibited the herds' ability to reproduce.[42] There were also the eastern Indians and, of course, the Texas settlers pouring into the southern tallgrass prairies, whose mere presence, much less hunting activities, certainly pressured the herds to the west. After 1840, a peace agreement between the southern plains nomadic tribes and Cheyennes and Arapahoes of the central plains erased a "neutral zone," where the herds had been able to replenish their numbers while Native hunters avoided the region for fear of bumping into an enemy hunting party.[43] By the 1850s, such circumstances encouraged the bison to move east into the only remaining neutral zone: north of the Arkansas River, between the semisedentary agricultural tribes of the Missouri River basin and the high plains tribes west of the 101st meridian.[44] These factors, combined with the destruction of river-valley ecosystems along the major trails by the gold rush travelers, traders, and Indian horse herds, constituted a relentless force pushing the bison to migrate east or perish.

Though such factors all contributed to the movement of the bison herds, the ecological effects of drought took center stage in dictating the migration of the buffalo. During dry years, plains grasses economize their use of water by using most of their resources to extend their root systems, tapping every obtainable H_2O molecule in the soil.[45] If this effort falls short, the grasses become dormant or even die off, thereby exposing the soil to sun and wind and allowing more resilient cacti to expand into the weakened areas. There is no doubt that large grazing mammals would find such conditions intolerable and would migrate toward regions with more grass cover and water.

The presence of a severe drought in the southern plains during those years explains why there were reports of buffalo so far east in the 1820s, and why Latrobe complained of not encountering any herds of bison as he crossed the prairies in 1832. It is likely that the herds moved back to the high plains in the 1830s and early 1840s because the short-grass prairies would have been able to sustain the large bison population in those fairly wet years. In Texas during the late 1840s and early 1850s, the herds moved towards the settlements that supposedly had been pushing them ever westward, and the subsequent movement of the Comanches to follow the herds strained Native relations with the Texans.[46] George Ruxton noted that the region from the Rocky Mountains to one hundred miles down the Arkansas River had been abandoned by the buffalo as well, but that he had seen them further east.[47] This movement corresponds with the spread of drought on the high plains from 1846 through 1848.[48]

In the early 1850s, the herds were again reported on the high plains, as the region just to the east was hit by two years of drought. In 1850 Joseph Smith, a nineteen-year-old boy from Buffalo, New York, accompanied a survey unit ordered to map the boundary between the Cherokee and Creek lands. Smith's diary is full of expectations of adventure, buffalo, and Indians. Yet what he found was a dry, hot grassland. The party was forced to drink from muddy puddles of water while following the course of the Arkansas River in July.[49] In fact, it was so dry that the lack of water threatened to terminate the survey and force the party to return to Fort Gibson.[50] Lieutenant Henry Heth, an officer at Fort Atkinson, located near present-day Dodge City, Kansas, wrote that from August 1850 to August 1851, it had not rained in any noteworthy amount within one hundred miles of the post, leaving very little grass for the animals to eat. With the exception of twelve mules and twelve horses, the draft animals were sent to Fort Leavenworth to alleviate the expense of freighting hay across the prairie to feed the stock.[51]

That year, Thomas Fitzpatrick, the Indian agent for the Upper Arkansas agency, wrote that his journey west to Fort Atkinson was marked by "a very unusual scarcity of water." When his group came upon the Arkansas River they found two small, stagnant pools emitting an extremely offensive smell due to the number of dead fish in the surrounding dry riverbed.[52] Charles Halleck, a visitor to Fort Atkinson in 1852, noticed the dearth of bison by the absence of animal droppings and the effect this had on gathering fuel for his party's fire pit: "'buffalo chips' . . . once found in great abundance, are now quite scarce."[53] But Fitzpatrick claimed that same year that buffalo abounded miles away, in the region between the Arkansas River and Fort Laramie.[54]

The severe droughts of the 1850s hit just as the major commerce trails began to chew away at the Canadian and Arkansas River valley grasses; but even without the presence of the Santa Fe and California Trails, the drought would have pushed the buffalo eastward. Because of their ability to adapt to climatic factors, the herds moved unexpectedly toward the rapidly settled Kansas frontier during the mid-1850s. Then, during the 1870s, years of average or higher rainfall on the short-grass prairie allowed the buffalo to move south to Fort Griffin in today's Texas, making the outpost the outfitting headquarters for buffalo hunters at that time.

This series of droughts both disrupted the southern plains trade system and increased violent conflict in the region. The Plains nomads could not procure enough buffalo hides to barter for the supplies of corn, guns, and other industrial goods they had become dependent on. The agricultural tribes did not have enough corn to trade with the Plains nomads for buffalo hides, making it more difficult to trade with the Anglo neighbors from whom they procured industrial goods. The presence of the removed eastern tribes prevented Plains tribes from migrating east to escape the drought, locate water, or follow the bison herds. These forces kept the Plains tribes starving on the withering grasslands during the drought with few options other than warfare or relying on the U.S. government for survival.

In the heart of the southern plains, the Comanches and Kiowas bore the brunt of the effects of the drought from 1846 to 1865. J. W. Whitefield, who was appointed to oversee the distribution of annuities and ensure tranquil relations between the United States and the Comanches, southern

Cheyennes, and Arapahoes of the Upper Arkansas agency, reported in 1854 that the Indians had begun stopping the passing wagons to beg for coffee and sugar, items for which they normally traded.[55] The next year, he mentioned that the Comanches had resorted to eating their horses and mules, truly an act of desperation for a people who measured status by the number of equines in their *remuda*. This desperation, Whitefield claimed, was the reason for their "frequent forays into Old and New Mexico."[56]

Some bands chose to accept reservation life and looked to the U.S. government for relief, a strategy for survival that would be replicated by non-Indian farmers during the drought of the 1930s. One Comanche leader, Tibbalo, asked his Muscogee neighbors to intercede on behalf of his people with the U.S. government. Tibbalo pointed out that "there were 5,000 of his tribe in a destitute condition" camped along the Arkansas River just west of the land granted to the removed eastern tribes. His people needed some of the land between the Arkansas and Red Rivers, which they had previously signed away.[57] The Comanches found themselves facing a new dilemma: they could not move east to follow the herds without going to war with the eastern tribes and offending the United States, yet to stay on the burned-up short-grass prairies was to starve. Many were even willing to take up agriculture and live on a reservation just west of the Muscogee tribe, if a treaty could be worked out.[58]

In Texas the southern Comanches found themselves in similar circumstances. During 1852 immediately following the drought years of 1850 and 1851, Texas's special Indian agent Horace Capron found the Comanches in a state of "extreme hunger bordering on starvation." Capron relates a speech given by headmen Ketumse and Sanaco eloquently describing the predicament of their people: "Over this vast country, where for centuries our ancestors roamed in undisputed possession, free and happy, what have we left? The game, our main dependence, is killed and driven off, and we are forced into the most sterile and barren portions to starve. We see nothing but extermination for us, and we await the result with stolid indifference."[59] By 1855, conditions had deteriorated to the point that these bands of Penateka Comanches were more than willing to accept life on the Clear Fork Reserve, a 20,000-acre area near Fort Belknap.

While some bands settled, other groups turned to raiding. The effect of drought on their horse herds was as catastrophic as the absence of buffalo to their way of life. The Comanches had been pivotal in the horse trade since their arrival on the southern plains in the early 1700s, but now that they relied on the equines for meat, they needed to find a source of horses and mules both for food and for trade. Mexico was the traditional target of Comanche raids. Following trails from the llano southward all the way to Zacatecas and Durango, Comanche war parties visited the rancheros yearly to take horses and mules, driving the livestock north to their homelands. The United States attempted to curb the raids, but the Kiowa and Comanche refused to end these annual forays.[60]

Ironically, the efforts of the U.S. Department of the Interior to discourage raiding into Mexico encouraged raids north of the Rio Grande. After 1854 Kiowa bands increasingly turned their attention on their western neighbors, the Navajo and Ute tribes. U.S. military authorities were able on occasion to turn the Kiowa war parties back, but then the farms and ranches of northeastern New Mexico had to endure the raids instead.[61] In 1857 Kiowa warriors raided the environs of Las Vegas and Moro, New Mexico. The following year, they invaded north-central New Mexico to obtain horses at the expense of the Utes.[62]

Raids increased on the eastern side of the southern plains as well. Douglass Cooper, Choctaw and Chickasaw agent, claimed that in 1857 "a state of war existed along the Choctaw and Texas frontiers." In 1858 residents around Fort Arbuckle in present-day southern Oklahoma reported the theft of more than seventy horses.[63] Comanche raids in Texas grew in frequency and intensity as well.[64]

During the years preceding and during the U.S. Civil War, the southern plains region was engulfed in its own form of internecine strife, as Plains Indians attacked neighbors on all sides with more frequency than ever before. Drought must bear a large portion of the blame for this eruption of warfare. Certainly, Plains nomads had long relied on raiding to supplement their horse herds, population, and trade items, but the 1850s seem unique for their excessive bloodletting. Plains tribes, in an attempt to reclaim their control of the hide trade, allied with each other, intent on driving all the prairie peoples out of the grasslands.

As the drought pushed the bison herds east towards non-Indian settlements, the largest herds relocated north of the Arkansas River and east of the one hundredth meridian. The common belief of that time, that the herds were being hunted into extinction in the southern plains, was only partially true. Hunting took its toll on the buffalo population, but drought had already made the buffalo's existence on the southern plains untenable.

Lydia C. Jones, who had moved as a young girl to Montague County, Texas, in 1855, remembered an encounter with thousands of the shaggy beasts. In her mind, the bison movements were tied to the availability of water. She stated in her "Reminiscences," "Red river was standing in holes then and [the buffalo] left the plains and came in [near Belcherville, Texas] to better range and water[. T]he soldiers began to think they would have to leave them on account of the dry weather and dust. When it began to rain the buffalo went back west."[65] Those living on the Texas frontier knew how important water was to their existence and associated the same dependence with the animals of the region.

In 1861, while Plains Indians were still very much involved in the hide trade, the bison returned to the region south of the Red River. John C. Irwin, a young boy when his family moved to the area near the Clear Fork Reserve, observed that "when we arrived at Camp Cooper there were a few buffalo in the country but after the soldiers left [for the Civil War] they came in by the thousands. I have seen them in their migrations both north and south, pass for a week at a time passing all day long by the thousands."[66] On April 21, 1866, the *Waco Register* reported that "the buffalo range 60 miles this side of Cooper and Phantom Hill, which has not been the case . . . for the last four or five years."[67] Although the buffalo returned to the southern plains, drought persisted in the tallgrass prairie of north-central Texas, creating circumstances that would prove dangerous to American Indians seeking the benefits of federal reservations.

FIVE

The Texas Indian Reserves

On the night after Christmas, 1858, a group of rugged Texas frontiers-men camped without a fire miles from their homes, braving the winter cold before daybreak. They planned a surprise attack on a band of seventeen sleeping Caddo Indians who had journeyed off their reservation. As the sun's rays pierced the horizon, the Texans attacked, charging the slumbering Natives' camp and firing into their pallets, killing some of the victims as they slept. Those that were able awoke to the smell of gunpowder and confusion as they grabbed for their weapons and made to escape. One young man managed to make it to his tepee's entrance before succumbing to the rifle fire of his attackers. Others slid under the sides of their dwellings as assailants blocked the flaps. The following day, military officials at the scene of the ambush reported four males and three females dead of gunshot wounds, and most of them in their bedrolls. All the other Natives were reported as injured from the fight.[1]

Violent acts were common on the western frontier as non-Indians and Native people meted out atrocities in a vicious cycle of violence. It was a dangerous time and place in which to live, the butchering of innocents an all too common affair; but, significantly, the attack of December 26, 1858, triggered the closing of the Texas Indian reserves, a little-known experiment in Texas history initiated to test whether Native people and non-Indians could coexist within restricted territorial boundaries. Although the odds were stacked against the endeavor's success, the outcome could have been much more positive if the new reservations had not been instated during a decade of widespread severe drought.

The Texas Indian reserves had their beginnings in the destitute camps of the Southern Comanches, whose hunting range had been diminished by the encroachment of non-Indian settlers on the Texas frontier. The pleas of some of these band leaders to receive a land reserve and become wards of the state, resulting from their inability to feed themselves during the prolonged drought, did not go unnoticed. Governor Elisha Pease fervently urged the state legislature to create Indian reserves, but the debate was not without its detractors, who asserted that Texas's unique sovereignty in retaining its public lands would be threatened by the mere existence of a federal Indian reservation.[2] Those favoring the creation of the reserves won the vote, and by early February 1854, the Texas legislature authorized the federal government to use up to twelve square leagues of unoccupied Texas land for Indian reservations. Uncharacteristically, the federal government did not dawdle, and within three months Captain R. B. Marcy led a survey expedition to identify suitable land. He was accompanied by a man whose future would be consumed with managing, protecting, and ultimately closing the reserves, Robert Simpson Neighbors.

Neighbors was born in Virginia in 1815 and became an orphan at the tender age of four months, when both of his parents died of smallpox. Fortune smiled on the young boy as Virginia farmer Samuel Hamner took up guardianship and provided a sound education for the young lad. At nineteen, the lure of adventure and fresh opportunities drew Neighbors west, first to New Orleans and then to Texas during its war for independence from Mexico. He arrived just as the conflict was ending, but served in the Texas Army from 1839 to 1841 as a quartermaster. Among the many colorful people he rubbed elbows with was famous Texas Ranger John C. Hays. Neighbors served in Hays's Ranger unit and on September 11, 1842, was in San Antonio attending court, which proved to be a classic example of being in the wrong place at the wrong time.

Texas president Mirabeau Bonaparte Lamar, a man whose ego matched his name, had claimed all of the land north and east of the Rio Grande River as part of the sovereign territory of the Republic of Texas. Whether as a ploy to open up a trade route which might eventually steal commerce from the Santa Fe Trail, or as part of his scheme to expand the empire of Texas, in June 1841 Lamar ordered a party of 320 men and 21 wagons loaded with trade goods to cross the high plains and negotiate the annexation of

Santa Fe. This excursion came to be known as the Santa Fe Expedition, and although it began with all of the optimism of a trip to an exotic land, its failure was titanic. After struggling across the trackless plains with all of those cumbersome wagons, the party had trouble finding water, and upon finally reaching Santa Fe, found the locals adamantly opposed to joining the aggressive fledgling republic. Mexican authorities arrested the members of the expedition and promptly marched them off to Perote Prison, near Mexico City.

General Adrian Woll, a French-born mercenary in the employ of the Mexican government, marched at the head of an expedition ordered to punish Texas for its arrogance and remind the "autonomous region" that its parent nation, Mexico, could discipline it at will. The invasion occurred during the presidential administration of Sam Houston, who strongly desired good relations with Mexico. So, as twenty-seven-year-old Robert Neighbors sat in court on September 11, 1842, the Mexican army marched into San Antonio. He became one of the 150 prisoners General Woll took back to Mexico and spent the next year and a half in Perote Prison, no doubt contemplating the arbitrary chance of timing.

Upon Neighbors's release, he returned to Texas, where President Sam Houston, busy reversing the policies of his political adversary, Mirameau B. Lamar, called on him to serve as Indian agent for the Tonkawas and Lipan Apaches. Here, the inventive twenty-nine-year-old agent abandoned the traditional method of maintaining a central headquarters where Native people could bring their grievances, and instead headed out into the frontier to visit the Natives himself. It was on one of these trips to the Tonkawa village that he met the Comanche headman Old Owl.

The Comanches and the Tonkawas were age-old enemies, and in 1845 Old Owl arrogantly rode into the Tonkawa village with forty or so warriors, demanding food and women. Neighbors, seizing the opportunity to make contact with the hostile Comanches, met the chief and mentioned that as agent for all the Indians in the region, he would like to instruct Old Owl on the benefits of "civilization." The Comanche leader, sporting with the white man, complimented his blue coat. Neighbors did not miss the insinuation and promptly took the coat off to offer it to Old Owl. Other warriors soon praised Neighbors's shirt, pants, and shoes until he was left standing in his underclothes. The Comanche headman was impressed with the impassivity of the white agent and invited him to accompany his warriors on a raiding excursion into Mexico. Neighbors, realizing this was the opportunity of a lifetime, accepted the invitation, and the bemused Comanche leader wryly claimed that instead of Neighbors molding Old Owl into a "civilized" man, he would make a horse thief out of Neighbors. The war party traveled south into Mexico looking for horses and captives to ransom or to add to the tribe. Old Owl planned an attack on a promising hacienda, but Neighbors, wishing to avoid bloodshed, called upon the targeted ranchero and offered to purchase two beeves on credit to appease the war party. The proud cattleman refused. Unimpressed by Neighbors's gesture, Old Owl then threatened to burn down the hacienda and kill everyone in it. The now not-so-cocky rancher prudently decided discretion was the better part of valor and provided two steers for the Indians' enjoyment.[3] These experiences were instrumental in Neighbors's development as Indian agent and friend of the Native peoples of Texas. Old Owl adopted him into the tribe, and as a result Neighbors was able to travel freely through the Comancheria, something few non-Comanches could boast of.

As Indian agent, Neighbors successfully negotiated two treaties: one between the Comanche people and the United States in 1846, and another involving the Comanches and the Texas Germans in 1847, and was also influential in mapping a transportation route between Austin and El Paso in 1849. Neighbors's accomplishments led to his appointment as Texas commissioner in 1849. In this capacity, he followed orders to organize counties in the eastern region of the newly annexed New Mexico territory, as many Texans concurred with Lamar that all the territory north and east of the Rio Grande was part of their grand state. He failed to convince the people of Santa Fe to agree to form a Texas county, but did organize El Paso County for the state. As a member of Texas's fourth legislature, he was one of the major advocates for Texas Indian reserves. Thus, he was the logical choice for Congress to send with Marcy on the expedition to find suitable lands for Texas's Indian reserves. Neighbors was well-acquainted with the rolling plains of Texas, and he knew that water was the key to successful settlements. With this in mind, he chose a site with springs and a picturesque, perennial river, aptly named Clear Fork, which could provide the sustenance of life year-round.[4] But the wisdom of his choice came to haunt the reserve Indians during the drought, as non-Indian neighbors desired the land adjacent to dependable water.

Knowing of the animosity between the Comanche and the Tonkawa, Caddo, and other eastern Texas tribes, Neighbors proposed and organized the creation of two reserves on the chosen land: the Brazos Reserve, 37,000 acres intended for the Tonkawa, Caddo, Waco, and Anadarko Indians near present-day Graham, Texas, and the Comanche Reserve, 18,500 acres along the Clear Fork of the Brazos, about forty miles west of the Brazos Reserve. The U.S. Congress earmarked $86,000 for the settlement of the Texas Indians, presenting economic opportunities for area cattle farmers, construction contractors, freighters, and others, especially as agency buildings needed to be built, food procured, and supplies shipped regularly to the reserves.[5] Some area ranchers, typically those without government contracts, had a differing opinion on the presence of federally subsidized Indian tribes in their backyards. Those settlers had been in a decade-long struggle to wrest land from the control of Native people, and tended to view all Indians as thieves, nuisances, and worse. It didn't help that the Comanches had preyed upon frontier ranches with ruthless abandon, stealing horses and capturing the settlers they did not kill. These non-Indian residents of the frontier felt threatened by the nearby presence of their erstwhile enemies.

Regardless of the local sentiments concerning the presence of the Indian reserves, the federal government continued their organization, appointing Neighbors to supervise the twin locations. Each reserve had its own agent: John R. Baylor at the Comanche Reserve and at the Brazos Reserve, Shapely P. Ross, father of famous Texas governor Lawrence "Sul" Ross.[6]

The beleaguered Caddo, Anadarko, and Tonkawa people no doubt were relieved to finally have a home and protection from both the warlike Plains Indians to the west and the growing non-Indian population, which had year by year encroached on their tribal land holdings. The southern Comanche bands led by Ketumse, Sanaco, and Buffalo Hump must have likewise been pleased to have a steady source of food in government annuities. Foreshadowing the difficulties the reserves would experience, the Plains reserve tribes suffered a setback as they gathered to move to the reservation in January of 1855, when a trader named Leyendecker spread a rumor among the tribesmen that certain white people were organizing a raiding party to annihilate the Natives. The rumor did have some basis in fact. Major General Persifor F. Smith had planned to take a military force out to punish hostile Comanches, but had said he would not harm the reservation Indians. Whatever the motivations of Leyendecker, he knew his targets well. Just fifteen years earlier, the southern Comanches had suffered the murder of thirty-five and capture of another twenty-seven of their people during an earlier attempt to create a peaceful coexistence between the Texans and the Comanches. That was in 1840, when a party of more than fifty

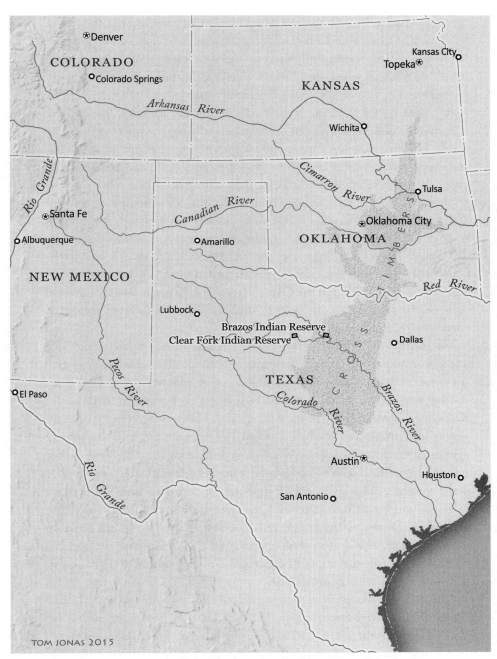

The Southern Plains. Map by Tom Jonas. Copyright © 2016 University of Oklahoma Press.

Penatekas had journeyed to the Council House in San Antonio to trade a female white captive and a boy of Mexican lineage for food and a guarantee of peace. The Texans, infuriated by the appearance of sixteen-year-old Matilda Lockhart, whose body the Natives had burned and disfigured, attacked the Indians, precipitating a wave of violence in retaliation. Could such a breach of a truce occur again? In the minds of many Comanches, it certainly could, and eight hundred terrified tribal members fled to the north. Ketumse was only able to hold together around two hundred of his followers to continue to the Clear Fork Reserve, while the smaller Texas tribes flocked without incident to the Brazos Reserve.[7]

By the spring of 1855, the Brazos Reserve was populated by Anadarko, Caddo, Ioni, Waco, Tawakoni, Delaware, and Shawnee Natives who lived in four nucleated clusters: the Caddo village, the Anadarko village, the Waco and Tawakoni village, and the Shawnee village, each near reliable sources of water. The Tonkawas arrived later and established their own village west of the others, and the Comanches under Ketumse lived for a short time on the Brazos Reserve, before Neighbors ordered their relocation to their own reservation farther west.

The formation of the reserves brought desirable jobs to the region. In addition to the agents, each location employed an interpreter, agricultural instructor, teacher, sutler, blacksmith, carpenter, physician, and day laborers; the Brazos Reserve also employed a minister. These jobs infused much-needed legal tender into the local economy.[8] The presence of government agents and employees sparked a boom in construction, and the Brazos agency alone erected seven single and two double log houses, along with a schoolhouse. The reserves also fostered a market for local stock, as the federal government provided 1,000,000 pounds of beef in the first year alone.[9] With the southern Comanches on a reservation and new forts in the vicinity, real estate speculation temporarily boomed.[10] These economic benefits won over many non-Indian settlers who lived close enough to the reserves to benefit from the government contracts, especially as the drought of the mid-nineteenth century intensified.

The growing drought hit the reserves as hard as the local farmers. The Caddo Natives were familiar with the agricultural environment east of the Cross Timbers, where they had successfully farmed for centuries, but relocated west, the age-old agriculturalists struggled to sow a crop. Although the Natives of the Brazos Reserve had planted 395 acres the first year, the "extremely dry season" reported by Superintendent Neighbors reduced the yields drastically, providing only enough corn to last into the fall.[11] The next year brought sustained drought in the region. On September 18, 1856, Neighbors wrote that "there has been an almost entire failure in the corn crops in Texas, from the grasshoppers in the spring and the extreme drought during the months of June, July, and August."[12] In addition to retarding agriculture, this drought triggered Indian raids on the frontier that increased as the drought intensified from 1854 to 1859. Opportunistic Plains warriors targeted the Brazos Reserve as well as non-Indian ranches, regardless of the presence of nearby Camp Cooper and Fort Griffin.

Comanche raids created near panic along the Texas frontier in the spring of 1855. Neighbors initiated a policy of confining the agency Indians to the reservation in an effort to flush out any who were taking part in the raids. He was particularly determined to shield the peaceful Natives from white reprisals, but he could not protect reserve Indians who strayed beyond the limits of the reserve.[13] In mid-September, word got back to the Brazos Reserve that the northern Comanches had declared war on all inhabitants south of the Red River, both Natives and white settlers alike.[14]

The Texas legislature ordered the establishment of Camp Cooper near the Comanche Reserve during the winter of 1856 to protect the Indian reservations and area settlers, and appointed Lieutenant Colonel Robert E. Lee commander of the post from April 9, 1856, to July 22, 1857. Due to administrative oversight, the post suffered without supply for two months. As officers attempted to bolster their food supply with fresh vegetables from newly planted gardens, they felt the effects of the drought. Lee wrote to his wife: "The weather still continues hot and dry with no prospect of rain, and our hopes for a few cabbage plants and roasting ears have passed away. We must bear it. The fact is that the Clear Fork no longer deserves its title but is converted into a stagnant pool."[15]

Newly arrived Comanche Reserve agent John Baylor claimed the cavalry horses at Camp Cooper were in pitiful condition: "The horses are poor and dying at an awful rate with the blind staggers." The blind staggers is usually a sign of selenium toxicity occurring from the ingestion of too much of that element. Selenium is found in soils formed from the breakdown of shale, often in arid climates. With a shortage of imported fodder, the horses of the post ingested much more of the local grasses and with them, the high levels of selenium.[16]

John R. Baylor was born in Paris, Kentucky, on July 27, 1822, to a U.S. Army surgeon. Such was the pull of adventure and opportunity in Texas that as a robust fourteen-year-old, he ran off from school in Cincinnati to fight in Texas's war for independence. Not far from home, he came across an older man who convinced him that he needed an education to truly profit from any adventures he might experience in that exotic frontier. He returned to school only to endure the barbs of his teacher, who brought the young runaway to the front of the class and exclaimed, "General Baylor and his army returned from the conquest of Mexico!"[17] Baylor later traveled with his mother to Indian Territory, where she took a job as head of the officer's mess at Fort Gibson while he opened a school for the Creek Indians. While there, he befriended the Creek Indian agent James Dawson, who had fallen into financial problems. In 1844 the agent was involved in an altercation with the licensed Indian trader, one Seaborn Hill, in which Dawson shot and killed the merchant. Apparently, Baylor was somehow involved, for he fled with Dawson to Texas with a $750 price on his head, while Dawson garnered a more substantial $1,500 reward status.[18] Baylor was ingenious in his ability to reinvent himself and no doubt used his connections to win the appointment as Indian agent to the Comanche Reserve.

Baylor began his duties as Indian agent for the Comanche Reserve well enough. He proposed growing wheat there instead of corn, which required much more water than wheat.[19] Neighbors followed his advice, and wheat became the preferred grain of the Comanches. The mounting violence on the frontier put Baylor in a tough predicament, however. Hostile Comanches continued to ride into Texas to steal horses, take captives, and garner military accolades, and some of these hostile Natives visited their kin on the reserves before and after their raids. Sometimes, as was common among Native societies, the victorious raider would leave a horse for his less fortunate reserve cousin. Thus, on occasion, settlers would track war parties to the Comanche Reserve and, to their chagrin, find their own horses grazing on the reservation grasses. To those settlers, who received no benefit from the reserves but who did see an immediate danger in their proximity to pioneering farmsteads, the situation was intolerable. The reserves and their agents came under fire from the growing non-Indian population. In an attempt to get to the bottom of the accusations, Baylor brought a Comanche woman who had just returned from a hunting foray off the reserve into his office. He suspected the party had committed some depredations while outside the boundaries of the reserve, and he locked the door behind her, cocked his revolver, and told the woman that if she failed to come clean

General John R. Baylor. 000-742-0014, William A. Keleher Pictorial Collection, Center for Southwest Research, University Libraries, University of New Mexico.

on where her people had obtained the new horses grazing near her lodge, he would kill her.[20] Although she confessed to Baylor that the men in her group had raided some ranches, Neighbors was not impressed by Baylor's method. Shortly following this incident, Neighbors took action against the Comanche agent for misappropriating reserve funds and dismissed him from the agency in 1857, creating a lifelong enemy.[21] Matthew Leeper replaced Baylor as agent on the Comanche Reserve.

The drought persisted regardless of the changes in reserve administration. High temperatures and dust storms were commonplace by the fourth year of the drought. One windy day at at Camp Cooper in May of 1857, Robert E. Lee had to pause from writing his journal to restake his tent pegs during a fierce dust storm. In his absence, the tempest blew so much dust into his shelter that his inkwell was clogged.[22] During the following summer, Neighbors was traveling with Comanche headman Ketumse when a sandstorm overturned two wagons, killing several oxen as the wagons blew away.[23] It appears that dirt storms are not the product of modern agricultural methods.

Although the wheat crop at the Brazos Reserve produced a successful harvest in June, Agent Ross gave the reserve Indians permission to leave the confines of the reservation to hunt.[24] His decision was not a popular one among area ranchers, but Ross did not understand the budding controversy over the reserves. The catalyst that ignited the northwest Texas frontier came in the person of John R. Baylor. The dismissed Baylor did not abandon the scene of his firing. Indeed, where else could he go? Texas was the ultimate location for starting over again. Instead, he put his education to use through editing a local publication, which he titled the *White Man*; in it, he blamed all the frontier's ills on the nearby reserve Indians. The venom he spewed in his hate journal was lapped up by its readers, who steadily grew in number and began clamoring for the closing of the reserves. By 1858, the increased Indian raids, cattle rustling, and horse thievery, all of which Baylor blamed on the Indians, had convinced many that the reserves were causing much more trouble than they were worth.

Hardin Runnels, ex-Speaker of the Texas House of Representatives and lieutenant governor, ran for governor against the "Hero of San Jacinto," Sam Houston, and he needed all the votes he could muster. Houston's pro-Union stance distanced some prominent people from his campaign, and Runnels tapped into the growing public concern about Indian raids by promising in his campaign to protect the frontier from "marauding savages." Many Texans considered Houston soft on Indians, while Runnels's tough stance on protecting the frontier from Indian depredations struck a chord with many. He was rewarded with the governorship.[25] In order to avoid incurring the wrath of those who opposed the presence of the Native people, Ross adjusted his policies for the Brazos Reserve the next year by confining his wards to the reservation.

There was no relief from the sun nor from the northern Comanches, who continued to raid the surrounding countryside as well as the Brazos Reserve. Non-Indian settlers claimed that Indian attacks along the Brazos and Colorado Rivers killed 7 non-Indian residents, took 600 horses, and caused $60,000 of damages in 1857 alone.[26] The return of severe drought conditions to the region that year caused a biting shortfall of supplies for basic subsistence.[27] In fact, for the fiscal year of 1857–1858, the federal government stepped in and provided 180,000 pounds of flour and 1,000,000 pounds of beef for the Natives of the Texas reserves, infusing currency into the local economy and providing avenues for profit by non-Indian ranchers, teamsters, and farmers of the region.[28] This surely angered those non-Indian settlers who did not have government contracts. In their minds, the Indians were savage raiders spreading death, destruction, and suffering, and all the while were subsidized by their own federal government! Baylor played on these sentiments in his publication, and the person he blamed the most for these injustices was his old boss, Robert Neighbors.[29]

In August 1858 settlers residing in Bosque County petitioned the secretary of the interior for the removal of Neighbors. This was the second such request from regional opponents of the reserves, and it would not be the last.[30] The accusations included in the petition were that Neighbors was allowing the reserve Indians to commit crimes against settlers and return to the reservation without retribution. The allegations were serious enough to warrant an investigation. That same year, the Interior Department sent Thomas T. Hawkins to Camp Cooper for five weeks to hear complaints about the reserve Indians and to visit the reservations and interview the people there. He reported a few meetings with settlers angry with the reserves, but in the end, he dismissed their charges.[31]

Even respected members of the Texas Rangers failed to find fault with the reserve Indians. As part of a campaign to stiffen frontier defense, Governor Runnels convinced the state legislature to appropriate $70,000 for the creation of a new, hundred-man Ranger unit to be headed by John Salmon Ford. Captain Ford and his men arrived at the Comanche Reservation in the spring of 1858 with the assumption that the reserve Indians were involved in the many frontier depredations. Interestingly, after a few months, the ranger officers found the reserve Indians innocent of the charges, and in their testimony claimed one of their own junior officers, Allison Nelson, had suggested they plant a trail from the site of an Indian raid back to the Comanche Reserve. Ford refused to go along with such a fraud, and through his discussion with Nelson, found that he harbored aspirations to become the next superintendent of the reserves after the removal of Neighbors.[32]

With the ever-increasing Indian raids on area settlers, the U.S. military instigated a new policy concerning hostile Southern Plains tribes. General David Twiggs, commander of the Department of Texas, announced an initiative to send military detachments into Indian Territory on punitive campaigns, taking the fight to the hostiles. Ford decided to beat the U.S. military to the glory and headed a Texas Ranger mission north of the Red River in May of 1858. Given the cloud of accusations coming from the pages of the *White Man,* Shapely Ross decided to prove the loyalty of Brazos Reserve Indians by recruiting some of the Native men to accompany the rangers to Indian Territory. Ross's Native warriors slightly outnumbered the ranger force and served pivotal roles as scouts, sources of geographical intelligence, and suppliers of food in hunting buffalo on the journey. Near the Antelope Hills on the Canadian River, the scouts discovered a trail leading to the hostile Nokoni Comanche camp of Iron Jacket. The Brazos Reserve Natives led the ensuing charge on the Comanche village and rounded up most of the 300 horses taken in the fight. The toll on the Nokoni Comanche was great, with a reported seventy-six of their number perishing in the battle, including their famed leader Iron Jacket. Upon the expedition's victorious return, Ford commented on the vital role played by the Brazos Reserve Indians and commended their support, but at the victory barbecues, local Texans ignored the contribution of the Natives while feting the rangers with praise.[33]

Even while Ford, Ross, and their men were away tracking hostile Comanches, the violence continued on the Texas frontier. Two non-Indian families were attacked and killed at their farmsteads a mere eighteen miles north of the reserve, and in the eyes of many settlers, this was obviously the work of marauding Indians.[34] Elsewhere, a party of white males came across an elderly Comanche man who was off of the reservation—all the excuse they needed to kill him and, tellingly, take his horses.[35]

To the north, the hostile Comanches replenished their vital lost stock of horses by raiding their eastern "civilized" Indian neighbors. Within a few weeks of Ford's attack on the Comanche village, sixty horses were taken from Chickasaw farms, and officers from Fort Arbuckle reported the theft of more than seventy mounts.[36] In August Comanche raids once again increased on the Texas frontier. The U.S. military was ready to commence its own invasion of Indian Territory in October, and Major Earl Van Dorn gladly accepted the aid of Ross and his Native militia to track and punish the bands that had committed the attacks. Ross found 125 reserve Indians willing to sign on to the expedition. Tragically, Van Dorn's force attacked a peaceful band of Buffalo Hump's Penatekas who had camped along Rush Creek on their way to an armistice conference at Fort Arbuckle. Although this act was clearly a terrible mistake, the Texans and their reserve Indian allies considered it a complete victory over their long time nemeses.

Still, the reserve Indians were caught between drought and the effects of northern Comanche depredations. That same year, Ross communicated his quandary to Neighbors: "From the 10th of June the Indians had no rain on their crops until they were entirely parched up by the drought. This fact and their being obliged to confine themselves to the limits of the reserve have caused a deficiency in the supplies required for subsistence."[37] Certainly these conditions, along with the seemingly more favorable view of the reserve Indians taken by settlers after their two "victories" in Indian Territory, had some bearing on Ross's decision to allow a few Anadarkos and Caddos, led by Choctaw Tom, to go south of the reservation boundary and graze their horses. In a report to Neighbors, Ross explained that Major Van Dorn was planning another offensive in the spring and had requested the aid of the Brazos Reserve volunteers; Choctaw Tom was planning to participate in Van Dorn's offensive but needed to build the strength of his horses for the campaign.[38] The lack of healthy grasses on the reserve due to the drought and confinement of the Natives' horses on limited pasture made outside grazing necessary. But the presence of reserve Indians away from the safe confines of the reservation caused fear and mistrust in those unsympathetic to the Native peoples' predicament. In an article for the *Dallas Herald,* John Sheen, a resident of Throckmorton County, described the suffering of the reserve Indians, but felt no pity: "During the latter part of the winter and early part of this spring, the Indians were in a starving condition, they annoyed me daily by begging and I was told went to every house in the neighborhood begging for something to eat, and said they were starving, and they looked as if they were. I have every reason to think they killed cattle in the neighborhood for the sake of getting something to eat." Sheen went on in the article to assert that reserve Indians stole some of his horses and that the agent paid him to maintain silence over the matter.[39]

The extended drought surely also had taken an economic toll on the Texas farmers and ranchers. During this time of economic crisis, the northern Comanche raids must have seemed all the more onerous. In times of distress, people look for scapegoats, and the accusatory fingers of many non-Indian settlers pointed to the reserves. In Texas, it was generally believed that the Wacos, Tawokonis, and Kichis were involved in raiding Texas settlements before they moved to the Texas reserves in 1855, and that they were obviously continuing these past activities.[40] Likewise, many settlers found it inconceivable that the reserve Comanches were totally innocent of involvement in their cousins' raids on Texas ranches. There were certainly non-Indian settlers who could profit from the removal of the Texas reserve Indians, as the reservations included some good farmland near perennial water sources, which becomes all the more desirable during repeated years of drought. The removal of the Texas Indians would open up that land to real estate speculation. Also, Baylor was still bent on his quest to destroy Neighbors and, now, any efforts to maintain the reservations. Often, people have a tendency to believe what they want to believe, and that was the case even in the 1850s.

It was in this poisonous atmosphere that Choctaw Tom and his group left the reserve to graze their horses. Some settlers told Tom that they did not want any Indians near their homes and that he and his followers had better return to the reserve or they would suffer the consequences. Tom agreed to return to the reserve but could not resist an invitation by another group of area settlers to hunt bear a little further downriver. On the night of December 26, a party of twenty men from Erath County waited near Choctaw Tom's camp and in the wee morning hours attacked, sneaking up to the teepees and firing point-blank into the bedding.[41] It is amazing that any of Choctaw Tom's group survived. Of the surviving twenty Native people in the camp, eight were wounded severely, and the others suffered some injury. Choctaw Tom survived, but his wife was killed and his daughter's thumb shot off.[42] Within days, a group of eight local settlers registered their disapproval of the massacre by signing a petition that claimed Choctaw Tom and his party were peaceable Indians who were wrongfully attacked.[43]

The massacre placed both the federal and state governments in an awkward position. More and more frontier settlers disagreed with both the national and Texas governments' policies that created and now protected the Texas reserves. Governor Runnels attempted to control the seething anger on the frontier by issuing a proclamation on January 10, 1859, warning non-Indian settlers against joining or assisting in attacks on the reserve Indians.[44] Then, in early February, the governor ordered Captain Ford, if he was "called upon by any peace Officer or other person deputed by Judicial authority," to "act promptly" to assist in the arrest of those charged with murdering the Brazos Indians.[45] District Judge N. W. Battle issued a writ for the arrest of those who participated in the attack, but still, no arrests were made.[46] For his part, Ranger captain John S. Ford refused to make any arrests, claiming he could only take orders for such action explicitly from Governor Runnels.[47] No such directive was forthcoming, as political necessity seemingly outweighed any desire to obtain justice for the reserve Indians.

Of course, the attackers initiated their own campaign to justify their actions. Eighteen of the vigilantes signed a sworn testimony claiming that they trailed to Choctaw Tom's camp a party who had killed a black man and stolen his horses, and that they then waited until daybreak to mete out frontier justice to the culprits. They regretted killing women and children, but felt that any Indians off the reserve were up to no good.[48] The two who did not sign the testimony were Samuel Stephens, who was killed in the attack, and John Barnes, who was recovering from wounds suffered during the butchery. John R. Baylor was not included in the group, but he certainly approved of their actions.

Neighbors, fearful of the growing animosity towards the reserves, had recommended the removal of the Texas reservation Natives to Indian Territory as early as the summer of 1857.[49] After the massacre of Choctaw Tom's hunting party, Neighbors redoubled his efforts and received permission from the commissioner of Indian affairs to begin relocating Texas Indians in the fall of 1859. The superintendent released this information to the public shortly after receiving the letter on March 30, but his opponents were not satisfied. Baylor's relentless smear campaign against Neighbors, evinced in his speeches and writings blaming area crimes on the reserve Indians, was winning the day. The ex-agent's charisma and convincing style of public speaking garnered many supporters, and groups of men began flocking to Palo Pinto County by early January 1859. Baylor and his supporters created a Jacksboro ranger unit with orders to kill on sight any Indian found off the reserve. In early May, members of this unofficial ranger unit discovered a small party of Caddo carrying official documents for Agent Ross. The "rangers" gave chase, captured, and later killed a courier named Fox, who had served admirably in both Captain Ford's and Major Van Dorn's punitive expeditions the previous summer. The murder of Fox so incensed U.S. Army lieutenant William Burnet, the son of interim Texas president David G. Burnet, that the captain took the surviving Caddos into Jacksboro under military escort to identify and arrest the perpetrators.[50] This impetuous act, though well meant, provided an inflammatory piece of propaganda for Baylor. He called for a vigilante invasion of the Brazos Reserve because U.S. authorities were obviously supporting the "Red Menace," even going so far as to threaten military force against law-abiding white settlers who opposed them.

On May 23, 1859, Baylor led a small army of around 250 men in an invasion of the Brazos Reserve. Captain J. B. Plummer of nearby Fort Belknap intercepted the vigilantes and ordered their withdrawal from the reserves. While Baylor returned to his force of supporters to contemplate his next move, his men captured and killed a Native woman and an eighty-year-old Native man. The vigilantes, proving their superiority and civilized manners, scalped the elderly victim.[51] When the Brazos Indians found out about the capture and murder of the woman and old man, they threw aside caution and sent a party of fifty warriors, of whom only thirty had firearms, to attack Baylor's vigilantes. The Brazos Indians sent Baylor's men fleeing to Marlin's ranch, two miles from agency headquarters. Baylor and most of his men took refuge in the ranch house, while the rest took cover in a nearby ravine. In the skirmish, two vigilantes were killed; two more died at Fort Belknap of wounds suffered at the ranch.[52] Captain Burnet, a witness to the day's activities, admitted in a letter to his father that "it seems a little hard to believe that 50 Indians should whip 250 white men who had come with the avowed intention of fighting, but it is nonetheless true."[53]

Governor Runnels appointed a peace commission of five well-respected frontier men to act as mediators among the "citizens of Texas," the "friendly Indians on the reservation," and the U.S. military stationed nearby.[54] The five men named in the appointment were George B. Erath, John Henry Brown, Richard Coke, J. M. Steiner, and J. M. Smith. The accomplishments of three of these men are worth noting. Erath was a veteran of San Jacinto, and he was instrumental as a state legislator in gaining the approval of the bill authorizing the creation of the new Ranger unit which was commanded by John Ford, who lived in the small village of Waco. Brown was a thirty-nine-year-old newspaper editor and member of the Texas Legislature who lived in Belton and who would later put his writing ability to use completing books on the history of Texas. Richard Coke was only thirty when Runnels tapped him for the peace commission. He resided in Waco as well and would later become governor of Texas. In a report to Governor Runnels, the commission sided with the vigilantes. Their statement claimed that the Brazos Reserve Natives had "killed cattle, and had horses at different times in their possession, under circumstances leaving no doubt of their guilt as horse thieves."[55] Recall that at least two of these men, Erath and Brown, were state congressmen who needed votes from the frontier to remain in office.

John Henry Brown, captain commanding one hundred state troops, followed instructions to keep the Brazos Indians within the limits of the reserve to prevent further bloodshed. The fact that the removal of the Texas reservation Indians was soon to occur and that many of the Natives' livestock were off the reserve did not matter to the one-time peace commissioner. He offered to detail men from his command to escort the Indians while retrieving their horses, but Neighbors dismissed the offer as hollow, having come to the conclusion that Brown, even as peace commissioner, was biased against the reserve Indians.[56] Brown was indeed in communication with Baylor and showed a willingness to accept his statements as fact. Investigating a rumor that Indians had burned Baylor's ranch house, Brown sent a detachment to the Comanche Reserve and then to Baylor's house, but found that no such vandalism had occurred. Baylor then informed the officer in charge of the Texas unit that the reserve Indians were stealing crops from area farms. Upon receiving this information, Brown ordered another detachment of Texas troops to the area near Baylor's farm with orders to attack any Indians found off the reserves and unaccompanied by a white man.[57] One Wichita Native left the reservation attempting to gather his stock before their departure and never returned. Local rancher and one-time Young County sheriff Patrick Murphy admitted that Brown had killed the Indian.[58]

In this atmosphere, Neighbors began organizing the relocation of the Texas reserve Indians to Indian Territory, where the Choctaw had agreed to give a large portion of their western domain, known as the Leased District, to the refugee Indians. He gathered appraisals for all of the Native people's property before their journey north and claimed $14,922.50 worth of livestock lost from the Comanche Reserve alone.[59] The Indian agent ordered supplies and wagons for the trip north, requested a U.S. military escort and on August 1, 1859, a Monday, 1,051 Natives left the Brazos Reserve while more than 300 Comanches departed from theirs. The effects of the drought were ever-present on the trek. Captain Burnet explained the circumstances to his father in a letter dated July 28, 1859, saying, "we will have nearly two thousand persons in the train and water is very scarce along the rout, I expect there will be much suffering for want of it: I never have seen such hot weather, as we have had here this summer: the Thermometer has averaged 106 [degrees] in the shade, from 10 O'C' in the morning until 5 O'C' in the evening, for the last month. . . . I expect many of the Women and Children will perish before we get there."[60] His prediction was prophetic, for even limiting travel to the morning hours to avoid the greatest heat, some Indians, ill prior to leaving the Texas reserves, perished from the arduous journey.[61]

Neighbors said goodbyes to his wards and friends amid the tears of Tonkawa headman Old Placedo. He hugged or shook hands with every Native present and some only reluctantly let go of him.[62] Matthew Leeper and others accompanied Neighbors back to Texas, where he hoped to negotiate the return of some of the stock the Indians had been forced to leave behind. Though on the return journey a mixed group of Indian and non-Indian livestock thieves attacked his party, seriously injuring Leeper, the group was back in Fort Belknap by September 14. That morning, the post commander warned Neighbors that his life was in danger, as various area residents talked openly of killing him, but the ex-superintendent disregarded these as so much bluff and exaggeration and traveled to the small town of Belknap, a half mile east of the fort. After spending two hours in the county and district clerks' office writing his closing reports as agent for the Texas reserves, he told the county clerk that he planned to return to San Antonio to be with his wife and children, "never to leave them again."[63] He left with a friend to return to Fort Belknap and check in on Matthew Leeper's condition when Patrick Murphy, the ex-sheriff who had confessed Brown's murder, accosted Neighbors with a gun, saying, "I understand that you said that I am a horse-thief. Is it so?" Neighbors warily denied the accusation, but placed his hand on his own pistol. Before he could get any more words out of his mouth, Murphy's brother-in-law and known gunman Edward Cornett stepped out from behind a chimney and shot the Indian agent with a double-barreled shotgun. Cornett escaped before the sheriff arrived.

Thus ended another tragic chapter of human affairs. The drought had created an environment in which reliable sources of water were vital to those wishing to remain in the Texas frontier. As settlers poured in to a region parched with a lack of rainfall, some gazed eagerly at the Indian Reserve lands and their perennial water supplies from the Brazos River and Clear Fork valleys. Led by the vituperations of John Baylor, the new settlers, who had no loyalties to the reserve Indians and who did not earn their livelihood through trade with the reserves, had fought for the removal of the tribes to

Indian Territory, hoping to make the region safer. The settlers of the Texas frontier would have been wise to reserve judgment on the guilt of the reservation Indians until all the facts were in. Lieutenant Burnet summed up the real effects of the reserves' closure: "Whether it will be for the interest of the people on the frontier remains to be proven—it will take a large am't of money out of the Country—will spoil a large market for beef and flour; and, deprive the Army . . . of their [the Indians'] services as Guides and trailers."[64] Further, Baylor's own son reflected years later that "the killing of Neighbors was about the greatest misfortune that could have befallen the frontier."[65] The raids, livestock theft, and killings continued; smarting from their treatment at the hands of the Texans, the reserve Indians would be much more willing to join the raiding parties headed south, and there was no one like Neighbors to calm the turbulent relations between settlers and Natives. Meanwhile, Baylor continued to pit the dueling frontier cultures against each other. The old man who convinced young Baylor to return to school could not have foreseen the damage that could result from a good education.

SIX

The Retreating Frontier of the Mid-Century

The removal of the Texas Indian reserves in 1859 did nothing to lessen the danger of the Texas frontier. Indian war parties often chose full moon nights to conduct their raids, and on such evenings every noise brought apprehension to the settler. Even the absence of night sounds excited terror. Only the most tenacious families remained, and they entrenched themselves in isolated but defensible ranch houses, riding out to check on their cattle only with great trepidation. The whole region seemed overrun with hostile Indians, even with the presence of U.S. military forts. As a result, according to reports, close to 80 percent of the people who had once built homes west of the Cross Timbers had fled eastward.[1] It was a time of fear and loneliness, or so goes the common view of the Texas frontier at that time. Historians have traditionally depicted the era as one of increasing Indian raids, vigilantism, and crime in a region famous for such acts of violence, and often historians blame the resulting outmigration on the withdrawal of U.S. troops as the Civil War approached. As this interpretation goes, the increased violence caused the frontier to retreat fifty to one hundred miles east.[2]

The scholars who argue that the outmigration from the Texas frontier region was a result of U.S. military withdrawal evidence their claims with contemporary newspaper reports, military dispatches, reminiscences, and settlers' requests to the governor for protection. However, letters from settlers to the governor of Texas claimed the Indian raids were causing depopulation prior to the removal of U.S. troops. Mr. W. Jones of Lampasas County wrote on October 17, 1858, that he had met with "several Reliable citizens Runing from the frontier with there familys and if protection is not amediately had that hole line will be avacuated."[3] H. Ryan, also of Lampasas, wrote, "Today I arrived home and found our people much more alarmed than when I left—large numbers have moved on into the counties below this, a number of families have been pursuaded to stop here and every house is full—all free of rent. We understand that what families are above this are forted up, None daring to attend to their busines."[4] In support of this traditional interpretation, census evidence clearly reveals a receding frontier. Wise County's 1860 population totaled 3,160 persons, but by 1870 the census data registered only 1,450 people.[5]

Yet, as noted historian Randolph Campbell observed in *Gone to Texas,* though settlers suffered from Indian attacks, such hostilities had "occurred before the war, and would continue afterward."[6] Both periods prior to and following the Civil War were eras of increasing population in the region. Showing incredible resilience and determination in forging a life on the frontier, pioneering families from the United States had been defending their homes from Indian raids for thirty years or, in the case of the Hispanic settlers, much longer, and the efforts of the U.S. military to deter the Indian forays had become a joke to the local citizenry.[7] Thus, even with the increased hostilities, the cause for the retreat of the frontier line some fifty to one hundred miles back east cannot be blamed solely on Indian raids, which the settlers had become accustomed to over the years.

Perhaps the answer to the question of true blame for frontier retreat is as close as a window. We need only turn our gaze outside the four walls of our offices and homes and look to the environment for some answers. Droughts have long been the cause for population migrations, most recognizably evidenced by the Dust Bowl of the 1930s. In a similar way, the repercussions of the severe drought of 1854–1865 incited many to leave the Texas frontier. Extreme aridity wreaked economic havoc on those relying on agricultural produce for their livelihood and pressured non-Indian settlers to find alternative sources of income, which often included moving away from home toward economic opportunity.

This drought of the mid-1800s was a vague memory by the latter part of that century. William C. Holden, a scholar who had lived through a few dry spells of his own, wrote about the impact of aridity on the history of Texas in an article titled "West Texas Drouths," published in 1928, that revealed through the recollections of old-timers that even thirty years later, the drought of 1864 was more memorable than the one of 1886–1887.[8] The vagueness of these memories did, however, keep the event from etching itself into the general memory of Texas history, allowing the more severe drought of the mid-1860s to slip into obscurity.

People living in the 1850s did describe the extremely dry conditions: A. J. Nixon and between fifty and sixty emigrants heading for California on the Old California Trail in 1856 camped at Big Spring, Texas, and while there decided to reappraise their goals. Their deliberations culminated in the splitting of the party, with one group heading south for Mexico and the other choosing to remain and settle in Gillespie County, Texas. The reasons they gave for foregoing the journey west included concern about Indian raids and scarcity of water—issues that also plagued the military.[9]

More than simply reducing water sources, the drought prevented the military from finding enough forage to maintain its horses' strength. Repeatedly, military officers had to call off engagements with hostile Indians because the cavalry's mounts were not fit to pursue the Native warriors, due in part to the absence of graze.[10] By 1860, the drought had denuded large portions of the frontier of grass cover. Cavalry officers were well aware of the hardship this placed on their animals, but still they preferred eastern stock that found the native grasses' nutrients less than sufficient. As a result, the military purchased corn to supplement their horse feed, which added the quandary of transporting feed while pursuing hostile Indians.[11] The impact of reduced grass cover and its subsequent effect upon the mounts of the Texas Rangers, both due to the drought, was devastating. John Williams wrote to Governor Sam Houston from Cherokee Creek in San Saba County that the grasses were not sufficient to feed the few horses they had left.[12] After only two weeks at Camp Cave, twenty-five miles northwest of San Saba, Texas, J. H. Conner reported to Governor Houston that, unless a heavy rain occurred soon, it would be necessary to change locations in order to find good grass and water.[13]

The severe dry spell would make campaigning difficult for the soldiers as well. In 1860, the U.S. military attempted a three-pronged invasion of Comanche territory moving out of Fort Riley, Kansas, Fort Arbuckle, Indian Territory, and Hatch's Ranch, New Mexico. Each unit had between two hundred fifty and three hundred troops in their command, which was a lot of canteens to fill. Officers called the campaign off, citing false information of the Comanche scouts and the drought.[14] Reminiscent of El Turco's attempt to devastate Coronado's Spanish army under the unforgiving sun and merciless wind of the Llano Estacado, the Comanche scouts had sent the cavalry on false trails to dried-up playas and rivers where the only source of water flowed underground.

These failed attempts to punish the Comanche raiders led a growing number of Texas frontier settlers to become disenchanted with federal efforts to protect them from marauding Indians. Recall that in the Reservation War of 1859, some claimed that the U.S. military protected the reserves in spite of threatened Texas citizens. Indeed, the U.S. Army had not been all that effective in keeping the hostile tribes at bay, and Texans had long placed more faith in their own Ranger units than in the national military. Thus, one of the main purported benefits of joining the United States, for military protection, seemed to many Texas settlers to be a misguided argument. Still, Sam Houston's defeat of Hiram Runnels in the Texas gubernatorial election of 1859 placed unionists in office.[15] In the following months, a significant event back east worked to undermine unionist support and propel the course of history in the Southwest, and indeed the whole nation, in a different direction.

On October 16, 1859, John Brown, a Massachusetts preacher turned ardent abolitionist, and seventeen accomplices crossed a railroad bridge into Harpers Ferry, Virginia, from Maryland on a foggy night, hoping to incite a slave insurrection at a federal arsenal there. The first victim of their mission was a black railroad worker who heard some shuffling on the bridge and hailed the group. Brown's party shot the worker dead to avoid discovery. From the bridge, they made their way to the federal arsenal in town and held the employees hostage while they waited for word of their mission to spread to the plantations of Virginia. They hoped a sizeable number of bondservants would flock to their banner, thus igniting a slave rebellion that could be armed by weapons from the arsenal. Instead, when the townspeople discovered the murdered railroad attendant and found the arsenal occupied, they surrounded the armory. The next day, Brown sent out two men to negotiate an exchange of prisoners, but the local militia made their intentions known by firing upon the intermediaries, initiating a fight that drove the conspirators to a nearby fire-engine house. That night, the secretary of war, John B. Floyd, organized a military response to the insurrection in Harpers Ferry, but Winfield Scott, the leading general in the

United States military, was absent from the capital. As it happened, Colonel Robert E. Lee was on leave from his post in Texas, visiting his sick wife in nearby Alexandria, Virginia. Floyd quickly notified Lee of the situation and sent him with a detachment of marines to pacify the situation. Lee and his command arrived in the afternoon of the 17th. The next morning, they stormed the fire-engine house, killing two of the conspirators and wounding Brown while suffering one death in their own unit.

Brown's plans failed miserably, but newspapers documented his time in jail from October 18 to his execution on December 2. He faced his death with dignity, claiming his life was worth more in his hanging than in any other purpose. To many northerners, he became a hero of sorts. Although many abolitionists denounced his actions, they claimed he had the correct intentions. To most southerners, to support a man who would encourage slaves to rise up and kill their sleeping owners was beyond reasonable explanation. The support for Brown among some abolitionists resulted in a growing southern distrust of abolitionists, Republicans, and northerners in general. Many southern whites lumped these three distinct groups into one cohesive mass, and a growing paranoia concerned with slave revolts swept through the southern states.

Political action, or more precisely, inaction, helped to convince many to join the secessionist movement in Texas. The U.S. House of Representatives was split between Democrats and Republicans, with neither holding a clear majority. As a result, the House was not able to appoint a Speaker from December 5, 1859 to February 1, 1860, and the deadlock would have continued if not for twenty congressmen from a third party known as the American Party who threw their support behind the Republicans simply to get things moving. Funding for frontier defense became a political turf war as Democrats pushed for it and Republicans delayed it. Secessionists viewed this as a breakdown in responsive government and held the Republican Party accountable for the damaging Indian raids.[16] John Baylor continued to use his education and penchant for invective in his Weatherford paper, the *White Man*. His editorials relentlessly attacked the Republican Party, calling for secession based on the federal government's failure to protect Texas citizens from Indian raids, and convincing many frontiersmen that secession was a better option than remaining in a Union dominated by Republicans.[17]

While tempers flared over secession, the environmental heat also increased. The clear skies of the severe drought continued to allow the sun's rays to hit the earth of the Southern Plains unimpeded, and temperatures rose accordingly. W. A. Sparks, a young man living near the border of Hopkins and Titus counties, Texas, reported a reading of 114 degrees Fahrenheit in the shade on July 8, 1860.[18] That Sunday afternoon, a trash fire started outside the W. W. Peak and Brothers store in Dallas. Within two hours every downtown building on the north and west sides of the square was reduced to ashen rubble. One and a half hours later, a worker discovered a fire in the counting room of James Smoot's store in downtown Denton. Soon the wind-driven embers from this fire ignited the roofs of neighboring stores. One of these establishments, Baines and Mounts, contained a twenty-five pound keg of gunpowder, which exploded, sending more embers and detritus high into the air. Nearly the entire western side of Denton's town square was destroyed. Others discovered fires in Pilot Point, Jefferson, Milford, Waxahachie, and Austin.[19]

A young editor for the *Dallas Herald*, Charles Pryor, wrote an article blaming these fires on abolitionists in a Harpers Ferry–like scheme to bring Texas to her knees. According to Pryor, two Negro slaves were arrested and "questioned," which could easily mean interrogated or perhaps even more accurately, tortured, for the burning of Crill Miller's house. They reportedly divulged a plot involving a group of abolitionist preachers whom local sheriffs had previously expelled from the state. According to the slaves' testimony, these preachers had returned to Texas and had organized these acts of arson in the hope that they could encourage a slave revolt on the upcoming election day. The confusion of coping with the fires might give their plan a better chance of success, or so the article claimed.[20]

While the festering debate over slavery came to a head, regional loyalty replaced what had been a strong Unionist sentiment in Texas. During Texas's years of independent status, the state took in many of immigrants from the upper South: Tennessee, Kentucky, and Missouri. Political heroes of the region, such as Andrew Jackson, were strong supporters of the Union and had opposed South Carolina's nullification efforts in 1832. Sam Houston represented Unionist Texans in their support of the United States, but by 1860 a majority of new Texans were coming from the Deep South, where no such sentimental attachments prevailed.[21]

With the election of Abraham Lincoln on November 6, 1860, Houston realized that secession in the Deep South was inevitable but hoped to forestall it in Texas, or at the very least keep Texas from joining the Confederacy. Seeing the tide of public opinion turning in favor of secession, he tried every maneuver he could think of to stall the political machinery. Legally, the Texas legislature had to call for a convention and for the election of delegates in order to proceed with secession, but the Texas legislature met biannually, and 1860 was an off year. Only the governor could call a special session of Congress, and Houston had no intention of giving secessionists an open door to pursue their agenda. On December 3, Oran Roberts, who had been appointed by Houston as Chief Justice of the Texas Supreme Court, called for a convention to vote on secession. His appeal, published in newspapers throughout the state and supported by prominent Texans such as John Ford, gained traction as county judges began organizing the election of delegates. Denouncing these acts as illegal but realizing the de facto truth that they would be perceived as legal, Houston called a special session of the state legislature for January 21, 1861, preempting the secession convention, which was scheduled for the 28th of that month.

Meanwhile, South Carolina had voted to secede from the Union on December 20, 1860, influencing some fence-sitters to leave their precarious perches and join the secessionist movement. Houston had hoped to persuade the Texas Congress to declare the elections and the convention illegal but, instead, it authorized the process, even offering its official chambers to the convention. Houston did convince his congress to demand that any convention decision regarding secession would have to be submitted to a popular referendum.

Secessionist forces hoped to obtain a break with the United States through valid elections and conventions, but Unionists questioned the legitimacy of secessionist actions. The county chief justices were supposed to prepare the delegates elections, but in some cases other local officials organized them. Intimidation at the polls influenced some Unionist voters to return home without having voted, while other Unionist voters refused to attend the election for fear their presence would lend the elections a much-needed validity. The results of the elections were sweeping victories for the secessionists. By January 26, five other Deep South states, Mississippi, Florida, Alabama, Georgia, and Louisiana, had voted to withdraw from the Union, further adding to the impetus for secession in Texas. On January 28, the first day of Texas's secession convention, the delegates elected Oran Roberts to preside over the meeting, leaving little doubt as to what the results would be.

Finally, on February 1, convention delegates voted 166 to 8 in favor of an ordinance of secession.[22] The convention went a step further by sending a delegation to Montgomery, Alabama, to join the Confederate States of

America. Governor Houston had tried to dissuade the convention from seceding by pointing out that such a course would result in the withdrawal of nearly three thousand U.S. military personnel, leaving the northern border unprotected from hostile Indian attacks. Seemingly in response to that argument, the convention superseded its authority by forming a Committee on Public Safety and granted that committee permission to remain in session so it could raise troops to defend state boundaries. The committee created three districts in the state and appointed commanders for each: John S. Ford received command of the southern district, including the military posts on the Rio Grande; Ben McCulloch, famed Indian fighter and commander of the Twin Sisters, six-pound guns made famous at the Battle of San Jacinto, was to command the central district between Forts Duncan and Chadbourne; and his brother, Henry McCulloch, headed the northern district. Another objective of the Committee on Public Safety was to organize the seizure of federal forts and arsenals within the state. As commander of the central district, Ben McCulloch began mustering troops to accomplish this goal.[23]

The commander of U.S. troops in Texas was Brigadier General David Twiggs. A native of Georgia, Twiggs had served honorably in the War of 1812, the Seminole Campaign, the Black Hawk Campaign, and the Mexican-American War. Now seventy years old and in poor health, Twiggs had been relieved of command by Robert E. Lee during a bout with illness, but had resumed his duties just after the election of Lincoln. As a southerner, Twiggs sympathized with the South but attempted to keep his military career untarnished while avoiding bloodshed. He repeatedly asked his superiors for instruction on how to handle Texas's request that federal forces evacuate the military posts in the state, but he received no answer. Worried about Twiggs's loyalties, Secretary of War Joseph Holt decided to honor Twiggs's request to be relieved of command in Texas before the evacuation of federal troops from Texas was resolved. Holt appointed New York native and ardent unionist Colonel Carlos Waite to the command, but before Waite could get to San Antonio to relieve Twiggs, McCulloch had forced the issue to a conclusion.

On February 16 around four o'clock in the morning, as Colonel Waite was making his way from Camp Verde to San Antonio to relieve Twiggs, Ben McCulloch ordered his men, numbering close to five hundred, to march into downtown San Antonio and occupy the plaza. Twiggs had issued orders to his garrison of 160 troops to avoid bloodshed in the case of an occupation by Texas irregulars. Thus McCulloch's move met with no resistance.[24]

During negotiations on February 18, McCulloch ordered all of the U.S. troops to withdraw from Texas soil without their weapons. Twiggs flatly refused to dishonor his men and himself, but offered to leave all military weapons, stores, and mounts behind, with the exception of personal arms. Nonetheless, the talks became heated when Texas commissioners found that Twiggs considered light batteries to be the personal arms of artillerists. Finally, Twiggs appeased the state's commissioners by accepting their demand that his troops leave by way of the coast rather than marching north to Kansas. The commissioners feared the presence of a ready army as close as Kansas and were willing to accept Twiggs's other conditions to avoid that.[25] On his way to Washington, D.C., to talk to General Scott about a promotion, Robert E. Lee entered San Antonio the day of the heated negotiations between McCulloch and Twiggs and found the streets full of men with red insignias on their clothing. Aware of the events leading up to secession and concerned over what role he should play in the event of a war between the states, he asked the proprietor of the Read House, where he was staying, what was going on. She replied that General Twiggs had surrendered to the Texans. Lee's next destination was the headquarters, where he hoped to discern the details of the capitulation. There members of the Texas Committee on Public Safety bluntly informed him that Texas was now a member of the Confederacy and pressed him to declare for the Confederacy as well, or they would withhold his personal effects until he did so. Fuming, he reproached the Texans and informed them that he was a Virginian, not a Texan, and his state had not severed its ties to the Union. Furthermore, he would make up his own mind about his actions. The Texans backed down, and Lee was allowed to proceed to Washington, where he declined command of the Union army after Virginia seceded, in order to defend his state and family.[26] Colonel Waite arrived in San Antonio the following day, hoping to rescind the agreement, but he found more than a thousand Texas troops in control of the streets and realized it was too late.

Henry McCulloch was also busy relieving federal troops of their commands. Having ridden 165 miles in six days, he and his two hundred or so troops bivouacked near Camp Colorado. After waiting a few days for reinforcements from Brownwood, his command approached the federal installation and he entered into negotiations with Captain Kirby Smith.

He made the same demands as had his brother, Ben, at San Antonio and received a similar reply from Smith who, although a southerner, refused to bring disgrace on his command and threatened to fight his way out of Texas, if need be.[27] Eventually, the two reached an agreement. Smith would allow a peaceful transition to Texan control of Camp Colorado, while federal troops could keep their arms, mounts, transportation, and supplies for ten days. In addition, when the U.S. servicemen arrived at the Texas coast, they would hand over all of the remaining mounts, wagons, and supplies to the state. But before this agreement could be recorded and signed, a messenger from General Twiggs arrived with news of the evacuation of San Antonio and orders for all U.S. forces in Texas to make their way to Galveston, retaining their personal arms. This order frustrated Henry McCulloch, who had negotiated better terms for evacuation of federal forces at Camp Colorado.

On February 23, 1861, Texans gave their consent to secede, voting 46,153 in favor to 14,747 against the resolution. Opposition to secession was most evident in the northern counties adjacent to the Red River that had been settled largely by migrants from the upper South and lower Midwest, and in the Texas Hill Country, dominated by German immigrants who owned few slaves and, as a minority in Texas, felt protected by the federal government.

Regardless of his intentions, Twiggs's surrender of nineteen U.S. posts provided $1,300,000 worth of war material to the Confederacy. Disregarding Twiggs's previous forty-eight years of faithful service, Secretary of War Holt labeled him a traitor and dismissed him from duty in the federal army on March 1, 1861. Colonel Waite and a few other federal soldiers failed to make it out of Texas before hearing the news of Fort Sumter's surrender on April 14 of that same year, and Confederate general Earl Van Dorn ordered the arrest of Waite and his men as prisoners of war. Even though Waite and the others eventually returned to the North, the poor treatment they received as prisoners created more hard feelings against Twiggs.[28]

The Civil War was approaching at the worst possible time for Texas. The combined effects of drought, Indian raids, and the funneling of men and resources east during the war would create truly hard times on the frontier. Just as Lieutenant William E. Burnet had predicted, Baylor's success in forcing the removal of the reserve Indians restricted the economic opportunities of area settlers. The withdrawal of the U.S. military devastated the incomes of many frontier residents as well. In 1855 Congress had discontinued hiring large eastern firms to freight supplies to the troops and instead, in an effort to reduce the cost of hauling, hired local freighters, who purchased supplies as close to the posts as they could.[29] In some instances, the local hires even amassed a modest wealth, as in the case of James Duff, a Scottish immigrant who held the position of post sutler at Fort Belknap from 1856 to 1859. His duties as sutler included procuring corn, hay, and flour for the post, as well as for Camp Colorado, Fort Chadbourne, and Camp Mason. By 1860 the entrepreneurial Duff had come to monopolize the supply contracts for most Texas forts.[30]

Like the reservations, military posts employed a wide spectrum of local people. Carpenters were needed to fix wagons and construct coffins; contracts for hay, corn, and oats attracted farmers to the nearby valleys; beef contracts kept many ranchers in business; and the construction of post buildings created a demand for lumber, triggering a growth in the lumber and, in turn, the sawmill industries. All necessitated the employment of a number of teamsters to haul materials and supplies to the posts, often at long distances. Of course, as today, military posts attracted practitioners of many other professions as well, including gamblers, bartenders, and prostitutes, as well as laundresses, physicians, and blacksmiths.[31]

The amount of money distributed throughout Texas by military spending was substantial. According to historian Thomas T. Smith, between 1849 and 1861, the United States spent $1,196,000 on forage, $191,000 on construction materials and labor, $61,000 for heating material like cord wood and charcoal, $30,000 on beef, and $140,100 on horses and mules for military posts on the Texas frontier.[32] No wonder Smith concludes that "the army was Texas' most dependable source of ready currency and the agency most responsible for the commercial development of the frontier."[33]

On a more local level, Camp Cooper, Fort Belknap, and Fort Chadbourne provided economic opportunities for a number of civilians and various settlers from the vicinity. The post commander at Camp Cooper in 1856 employed a guide, a clerk, six teamsters, four mail carriers, and a commissary operator. Fort Belknap likewise employed a smith, a laborer, two express men to Camp Cooper, two express men to Fort Washita, three herdsmen, four teamsters, and a mason; Fort Chadbourne employed the services of a clerk, four teamsters, two herdsmen, three mail carriers, and, occasionally, a carpenter and a smith. Supply officers procured fresh beef from local ranchers at a price of 6.7 cents per pound, and the inability of the camp garden to produce vegetables meant that most other items bound for the camp dinner tables had to be purchased as well.[34]

In 1857 Lieutenant T. A. Washington of the First Infantry, stationed at Fort Chadbourne, reported to Major D. H. Vinton, the principal quartermaster for the Department of Texas, that animal feed was delivered to the fort at a cost of $2.84 per bushel, and that he had contracted 4,000 bushels with local farmer Mr. I. C. Gooch. With regard to hay, he claimed "there is no hay on hand, can be had of a good quality from the vicinity of the post. The cost of hay delivered last year was $25.00 per ton, and was delivered by contract, and was cut in a valley some 2 miles distant. The grass up to present time is sparce." A contractor supplied beef at 9.5 cents per pound. The post obtained firewood from land rented out by Mr. S. A. Maverick for $25.00 a month and from land rented out by Messrs. Twohig and Howard for $50.00 a month. Contractors hauled construction lumber from Bastrop, Texas, for $110.00 per thousand board feet, and shingles cost $10.25 per thousand.[35]

Such was the economic impact of having the protection of a fort nearby that the Texas Emigration and Land Company used the news of an impending fort construction to lure settlers to their holdings.[36] Property values near Fort Belknap rose 200 percent when news reached the area that the post had become the headquarters for the Second Cavalry.[37] The presence of many military posts encouraged the growth of nearby towns, enabling them to survive the later removal of federal troops from the state.

These posts were not meant to be maintained forever, but similar to the military cutbacks of the early 1990s, closing a military base met with a lot of resistance. The United States established Camp Belknap in 1851 near present-day Newcastle, Texas, and it soon gained the status of a fort. Unfortunately for the local settlers, Fort Belknap was abandoned on February 17, 1859, due to lack of sufficient water.[38] The troops and supplies were removed to Camp Cooper, forty miles further west. General Twiggs, who ordered the closing of the fort, was deluged with requests to keep it open, and he suspected that most of these missives were from merchants, sutlers, bartenders, and brothel owners, whose pocketbooks suffered the most from troop removal.[39] Yet, with the coming of the Civil War, the federal forces withdrew, and most of the Texas posts were closed.

The removal of federal units from Texas between February 18 and April 13, 1861, certainly undermined the commercial activities of frontier settlers. Hostile tribes may well havetaken advantage of the situation and increased their raids into the Texas frontier. It is also probable that some frontier settlers used the Indian raids to convince the governor to create local militias to defend the region; militiamen would be rewarded with small monthly salaries that could compensate for revenue lost in the drought. Shortly after the closing of Fort Belknap, James W. C. Dechman wrote to Governor Sam Houston on May 16, 1860, to request the formation of a minute company of eighty to one hundred men, with pay for the minutemen in state funds and corn feed for their horses.[40]

The relocation of the troops from Fort Belknap and Dechman's request to the governor reveal a few things about the region. First, as previously noted, the drought had made it difficult to procure water in an area that had been recommended as a site for a military post specifically because it had ample water. Second, the various citizens who had enjoyed government contracts were now scrambling to bring in money. Dechman had lived in the region for three years, one year and three months of which had been without the presence of troops at Fort Belknap. It would seem that the fifteen months Dechman waited to notify the governor of the dangerous conditions of frontier life without military protection was also enough time for him to recognize the effects of having no market for his cattle or agricultural products. His request for corn feed reveals that he was finding it difficult to feed his horses. It is certain that the drought and federal troop removal played a major role in Dechman's plea for the organization of a minute company, which could also stimulate trade. The corn feed would have to be purchased somewhere, and the state funds dispersed to the men would bring currency into the region now devoid of a military base.

On December 5, 1860, another citizen of the frontier wrote to Governor Houston, echoing Dechman's concerns. W. W. O. Stanfield stated that "since the withdrawal of col. Johnston's force from the frontier the Indians are again amongst us."[41] One wonders if the Indians had ever failed to be among the settlers of the frontier. The years preceding the removal of U.S. troops had provided many examples of the military's inability to protect the frontier adequately. During the late 1850s, locals had blamed numerous Indian raids on the reserve Indians, only to find that after these peoples' relocation to Indian Territory, the raids continued. The influence of drought on frontier settlers was ever present, as citizens of Burnet County remarked in their "Petition From the Citizens of Burnet County to Sam Houston," sent to the state capital with twenty-three signatures on July 25, 1860. The petition underscores the growing hysteria concerning Indian raids, the economic constraints on the frontier, and the effects of drought:

Our County is again full of Indians, almost in every direction. Some Fifty Horses have been stolen, some killed and a great many missing, in a word Our People are greatly excited, and unless something can be speedily done our county is ruined. Aiming to the long protracted drought our only means of support for our families depend upon our stock, and if the people are left long in the present condition, they will be compelled to remove their stock at a great sacrifice and leave those behind that are unable to move in a destitute condition. We ask some relief from the state.[42]

Thus, the "rolling back of the Frontier" that has largely been attributed to increased Indian raids was most certainly also influenced by the prolonged drought. Out-migration has always accompanied extreme dry conditions, and the arid episode of the mid-1850s to mid-1860s was no exception. In 1854 John H. Chrisman had journeyed to Texas to survey the original townsite of Gatesville. After settling and working there for several years, Chrisman claimed that in 1857 "at least two-thirds of the settlers west of Gatesville left the area and went east with the hope of avoiding the drought and Indian raids."[43]

As people moved away from the region, the Indian raids must have become even more terrorizing, but the cloudless days of the sustained drought brought another kind of apprehension. With no rain, there could be no crops, no fodder for animals, and no economic opportunity for area farmers and ranchers. While the receding frontier was to some extent the effect of human actions in the form of Indian raids, it was primarily an environmental series of events that contributed to the depopulation of the Texas frontier during the 1860s. As severe drought crippled economic opportunity for agricultural production, settlers turned to livestock sales, but the removal of the Texas Indian reserves and federal military posts took away the livestock market. It is worth noting that if rainfall conditions had been average during this period, and the increased motivation for Indian raids were thereby removed, most settlers would have been able to remain on the frontier during the 1860s just as they had for decades before.

SEVEN

The Five Tribes and the Confederacy

I n their new Indian Territory tribal holdings, the removed southern tribes had attempted to reestablish lives as similar as possible to those they had left behind. Some of the wealthier members established plantations based on slave labor, just as they had done in their eastern homelands. Others continued to rely on their own skills as farmers, especially the wealthier people of the southern limits of Indian Territory. These people, acquainted with the weather of the more humid southeastern United States, were not familiar with the fickle climate of the grasslands. Fortunately for them, the period from 1830 to 1845 was wetter than usual, allowing them to subsist on their agricultural production as they had before. These families eventually established large plantations on the Red River and its tributaries and continued their production of cotton, but before this investment could become profitable, limitations on the Red River's waterborne traffic had to be overcome.

An immense logjam known as the Great Red River Raft stretched from Natchitoches, Louisiana, to near the mouth of the Kiamichi River in to-day's southeastern Oklahoma, a distance of 165 miles.[1] This obstruction to the river's flow backed up the water, creating swamplands. Thus, any river traffic would need to unload and bypass the logjam and its swamps via overland trails, then be reloaded onto barges or paddle boats bound for New Orleans. Work to clear the raft began in 1833 to facilitate the resupply of Forts Towson, Arbuckle, and Washita. After five years of difficult and meticulous work, the U.S. Army Corps of Engineers under the supervision of Captain Henry M. Shreve completed the herculean labor. When the Red River was finally opened to traffic in 1839, the main river-borne exports from neighboring Indian Territory were cotton, corn, and pecans.[2] The various military posts in Indian Territory also provided markets for a substantial amount of local produce.

During the wet era from the late 1820s to the mid-1840s, non-Indian travelers praised Indian Territory. In 1832 famed artist George Catlin, while accompanying the Richard Irving Dodge expedition from Fort Gibson to the mouth of the False Washita River, claimed that "this picturesque country of two hundred miles, over which we have passed, belongs to the Creek and Choctaw, and affords one of the richest and most desirable countries in the world for agricultural pursuits."[3] Thomas Farnham, a young Vermont lawyer who had moved to Peoria, Illinois, for health reasons, decided to move to Oregon country after listening to a missionary report on the marvelous condition of the land there. He departed in the spring of 1839 and took the Santa Fe Trail west to the Rocky Mountains and points beyond, though his report on the trip included many descriptions of land not quite on his path. Regardless of where he got his information, he was so impressed by the area south of the Canadian River and north of the Red River that he claimed "the country [was] capable of supporting a population as dense as that of England."[4] Indeed, the tribes of Indian Territory harvested abundant yields during this wet cycle. In 1833 the Choctaw corn crop is reported to have produced a surplus of 40,000 bushels, and this surplus increased to 50,000 bushels in 1836. That same year, 500 bales of cotton were exported down the newly unobstructed Red River.[5]

Given the level of commercial agriculture in Indian Territory, major droughts had disastrous consequences there. From 1850 to 1851, residents of the Choctaw Nation endured the effects of back-to-back years with below-average moisture. On October 15, 1851, the *Choctaw Intelligencer* of Doaksville, Choctaw Nation, reported that the Red River was at its lowest stage since the Choctaws had arrived in the region and that the closest any steamer could come to Indian Territory was Alexandria, Louisiana.[6] Those relying on river transportation to export their crops certainly became acquainted with the consequences of low water, but this drought lasted only two years. The oncoming drought would continue for ten years, with only one year of above-average rainfall.

The mega-drought of the mid-1850s began, as all droughts do, with a dry month that extended to two, three, and more. Local farmers must have thought the streak of dry weeks would break any day. but the conditions continued to prevail from June 15, 1854 to the following June, with little in the way of rainfall. Military installations often kept meticulous weather records. In 1842 the United States had established Fort Washita fifteen miles upriver from the confluence of the Washita and Red Rivers. Records from this installation report a scant 18.83 inches of rain during 1854, less than half of the 38.33-inch yearly average that the fort registered from 1844 to 1857.[7] The condition of the land was so poor that when a band of Choctaw immigrants from Mississippi arrived that year, they became disillusioned and decided to return east.[8] George W. Manypenny described the tribulations of the Five Tribes in his Report of the Commissioner of Indian Affairs for 1855: "The great drought of last year almost entirely destroyed their crops and subject[ed] them to much trial and suffering, which, however, they bore with commendable fortitude."[9]

This drought affected almost every facet of economic enterprise in the nation. Local cattle ranchers had already become involved in an extended commercial system. Indian cattlemen pushed herds up the East Shawnee Trail and delivered them to Missouri stockmen, who fattened the beeves on cornmeal before selling them to migrants on one of the trails along the Missouri River or exchanging them in Saint Louis for eventual butchering or resale.[10] By 1856 Indian Territory ranchers had sold most of their remaining cattle to California buyers who intended to drive the animals west to feed the thousands of people flocking to the gold deposits of the Sierra Nevada. Those who sold their cattle early on in the drought did not have to watch them perish in muddy, dried-up creek beds where small pools of water were churned into mudholes by cattle frantic for water, until even these feeble sources of hydration turned to cracked earth.[11] Due to the low production of the wells, troops at Fort Washita constructed cisterns to store water. This, of course, was not very effective during the drought. The only available water was more than two miles away, and it had to be hauled to the outpost daily.[12]

After a fairly wet November, in which 3.54 inches of rain accumulated in the gauge at Fort Washita, the winter months proved to be extremely dry, registering only 3.25 inches total from December to March. In response to the previous year's hardships, Indian agents encouraged the tribes to break more land, raise more crops, and store the excess in a reserve food supply in case it was needed. The drought had impressed this need on the agents as all reserves of corn and other seeds were consumed rather quickly in 1854, leaving no seed for planting the following year.[13]

Unfortunately, another year of excessively dry conditions was in store for the Choctaw Nation. As expected in drought years, the watercourses did not carry enough water to run, but by the second year of these conditions the springs began to fail as well. As rainwater penetrates the soil and follows the force of gravity along fissures and through porous rock below the earth's surface, it often comes to an opening on the earth's surface and trickles out into daylight as a spring. When, as in the drought year of 1855, there are no rains to replenish the underground water supply, the water table drops below the spring outlets. Simultaneously, the Red River was continually too low to allow waterborne traffic, precluding the cheapest and easiest means for transporting grain and other necessities to the towns of Indian Territory.[14] In the words of J. C. Robinson, the superintendent of the Chickasaw Manual Labor Academy, "we have had but very little rain in this section of the country since last June one year ago, now fourteen months. . . . Our streams are dried up, stock water gone, and our springs are failing, so that the prospect before us, in this respect, is gloomy indeed."[15]

The privation created by the lack of rainfall is brought to life in the letters of John R. Whaley, who, as a young Virginian down on his luck, signed up with the military for adventure and a steady paycheck. After a long boat transit from Baltimore to New Orleans, his troop was transported on a steamboat to Little Rock, Arkansas, where it disembarked on a foot march to Fort Washita. Private Whaley recounted the suffering he and the other men experienced from lack of water. They were forced to drink from stagnant pools and, to no one's surprise, battled bouts of diarrhea for the remainder of their journey, placing them in a deadly situation. Cramps and sweating dehydrated their bodies, but the only available water would be sure to worsen their condition. Whaley was one of the lucky ones, for he completed the journey all the way to Fort Washita. Five others were buried along the way.[16]

In 1855 an agricultural catastrophe approaching biblical proportions occurred. As in ancient Egypt, a plague of grasshoppers descended on the inhabitants of the Choctaw Nation to complement the severe drought and the suffering it induced. Insects swarmed over the crops of Indian Territory from spring through fall of that year.[17] The factors influencing grasshopper populations are myriad and finely balanced within the ecological framework of the grasslands. Reproductive rates among the insects are highest during dry and warm weather, averaging 78 degrees Fahrenheit or higher, so populations boom when the preceding fall has met these criteria, allowing a high harvest of eggs to be planted into the soil. The late summer and early fall months of 1854 were dry around Fort Washita, yielding only 0.59, 1.39, 1.17, and 1.19 inches of rainfall, respectively, from July through October.[18]

Warm and dry temperatures also inhibit predators and pathogens that could reduce the grasshopper population. Parasites and invertebrate predators, such as robber or asilid flies, wasps, ants, and spiders, thrive in humid conditions. The vertebrate predators that feed on grasshoppers, such as skunks, coyotes, badgers, bobcats, foxes, field mice, toads, snakes, and lizards, as well as horned larks, western meadowlarks, and lark buntings, also thrive in more humid climes. Finally, the most threatening organisms to grasshopper populations are pathogens such as fungal diseases and bacteria, both of which require humid conditions to thrive. The dry fall of 1854 placed stress on the predator populations and retarded the development of diseases among grasshoppers.

Of course, the availability of food sources is mandatory to support such a population boom. Studies have found that a mixed diet of legumes and nitrogen-consuming plants is most beneficial for grasshopper reproduction. Further, to maintain populations over the entire summer, there must be enough rainfall during the critical plant-growing season in the spring to ensure prolonged plant growth in the region.[19] Farmers in 1855 experienced adequate rainfall during the growing season; then the weather turned dry, lowering the annual rainfall to far below the region's average. The weather records at Fort Washita report adequate rainfalls for April, May, June, July, and August of 3.49, 2.11, 2.07, 2.39, and 3.54 inches, respectively.[20] High grasshopper mortality has been associated with rainfall, while high grasshopper reproduction has been correlated with warm, dry conditions. Yet successive years of extreme drought retards the growth of insect populations as well. Lack of rainfall can reduce vegetation for feed, and continued dry conditions compact the soils, prohibiting grasshopper nymphs from accessing the earth's surface and growing to maturity. Thus, such a boom in the grasshopper population requires ideal conditions with rainfall increasing and decreasing at just the optimum times.

The year 1855 presented just such ideal conditions. The warm and dry fall of 1854 encouraged high egg yields, and the absence of early and late freezes allowed the eggs to make it through to the spring, which also was fairly warm and dry. Adequate rains in May and June of 1855 gave the vegetative cover enough moisture to grow for the remainder of the summer and provide feed for the high insect populations. Then, the return to hot, dry weather in July and August permitted high reproductive rates, discouraging the presence of pathogens and predators that can check grasshopper populations.

The twin scourges of drought and grasshoppers wreaked havoc on the daily life of the Choctaw. Harvests were slight, and area farmers had already consumed the seed supplies, leaving little to plant for the upcoming year. The failure of the corn crop meant that there would be no feed for the hogs, the main source of meat during the antebellum era. Most of the region's cattle had been bought by California drovers and the arteries for importing supplies, the Arkansas and Red Rivers, were unreliable. These conditions drove market prices beyond the means of the average family. In the Choctaw Nation, many faced famine and turned to stealing from their wealthier neighbors for sustenance.[21]

Although 1856 was another year of drought, the suffering was not so widespread. To be sure, certain areas experienced complete loss of crops, but others were able to produce a fair harvest. C. W. Dean, the superintendent of the Indian Nations, acknowledged the drought as a "trial" that devastated crops from one end of his superintendency to the other but felt that the region would be able to supply "the wants of the people" with produce from the areas that had good harvests.[22] The following year brought ample rain to the Indian Territory. The only setback to a perfect harvest was a late freeze on April 6, 1857, in which the thermometer reading as far south as Eagletown in far southeastern Choctaw Nation was a mere 18 degrees Fahrenheit.[23] Frost did cut back the spring corn, wheat, and fruit crops, but bountiful later harvests were reported from the Choctaw agents.[24] In fact, the late summer and early fall months of September, August, and October registered 11.23 inches of rainfall at Fort Washita, more than a third of the accumulated 29.04 inches that year.[25]

Unfortunately, the drought resumed in 1857 and 1858. Elias Rector, the regional superintendent of Indian affairs, observed the typical reaction to drought on the plains and its periphery: migration. In his report to Commissioner of Indian Affairs Charles Mix, he wrote, "Most of the Cherokee, Creek, Seminole, Choctaw, and Chickasaw, cultivate the soil to a small extent; but having no individual proprietorship therein, they are continually on the wing, moving from place to place; and one sees, in traveling through their country, more deserted than inhabited houses."[26] Although Rector perceived common landholding as the cause for the desertion of homes, the more obvious contributor was drought. The Natives of the drought-stricken area left their homes out of necessity to obtain work, to live with families who were better off, or simply to search for water. In Kansas, the same type of environmental conditions produced a similar exodus among non-Indians.[27]

The spring of 1859 provided enough moisture to ensure a good harvest of wheat, which requires healthy rains in April and May and a fairly dry June, so that the heads can dry out and avoid rust. Such was the case during 1859, with a drought setting in after May 15 and running through late July, when the corn crop required rainfall to offset the rising temperatures. Most agencies reported good wheat crops, but extremely poor corn crops.[28] Corn was the main staple of the Choctaw diet, especially *tofulla,* a dish of corn soured in a boiling pot; *pashofa,* a mixture of meat and cooked cornmeal; *walusha,* cornmeal mixed with grape juice and sugar; *bahar,* a dough made of cornmeal, sugar, and pulped hickory nuts and walnuts; and *abundha,* cooked corn made into a dough, mixed with cooked beans, and boiled in corn shucks.[29] A poor corn harvest would greatly reduce the traditional meals of the Choctaws, forcing them to make do with substitute ingredients or find other dishes entirely. John Edwards, a Presbyterian missionary to the Choctaws during the pre–Civil War years, recalled how resourceful the Indians of the Nation were during the corn famine of the winter of 1860 and 1861. He claimed corn was so scarce that few families enjoyed any meals with the Choctaw staple. Instead they made do with acorn mush, acorn bread, and wild potatoes found in the swampy areas close to the Red River and other tributaries.[30]

Certainly 1860 proved to be one of the most severe years of drought in Indian Territory. The winter was cold and dry, and the spring continued uncharacteristically dry, though often the months from March to June are the wettest for the region. The Choctaw Nation bore the most intense drought of the area. The previous year had been one of hardship, but George Ainslie, the superintendent of the Koonsha Female Seminary in Goodwater, Choctaw Nation, claimed that "the prospects for the coming year [were] tenfold more gloomy."[31] Cyrus Kingsbury, a Congregationalist missionary from New Hampshire who worked among the Choctaws before their removal from Mississippi to Indian Territory, expressed concern for the poorer people of the Choctaw Nation, wondering how they could fend for themselves with their small farms and gardens completely burned up. The Choctaws altogether harvested only a fourth of their usual yield of all crops combined.[32]

Others wrote of the drought as well. Sue McBeth left her home in Fairfield, Iowa, in the spring of 1860, in response to an invitation to become a teacher at one of the Presbyterian Indian schools in the Choctaw Nation. She kept a journal that documents the impact of the prolonged drought on the people of southeastern Indian Territory. Her entry for August 1, 1860, discusses the difficulty of obtaining flour:

> I hear they have good crops in the states, but the difficulty is getting it here. Gaines Landing, the nearest point on the Mississippi is not less than 200 miles from here, and it would need to be brought here in wagons [implying the rivers were too low to allow water borne traffic], almost an impossibility under the broiling sun, for the heat contracts the wood of the wheels so that the tires fall off. Why Tawnee Tubbe went to Texas after a load for Father Kingsbury and his wagon fell to pieces and had to get repaired. That cost $12.00. One of his oxen fell dead. That cost $25.00 more, making $37.00 in all beside his time and labor.[33]

She also mentions a debate concerning how to address the suffering of the Choctaw people during the drought-induced famine. The missionary schools, including the one at which she worked, were funded partly through appropriations the U.S. government made to the Choctaws. In light of the incredible suffering, some thought this money could be put to better use feeding the starving people.[34]

Elias Rector urged the Commissioner of Indian Affairs, A. B. Greenwood, to recommend that the federal government immediately appropriate funds to pay the tribes part of the $2,981,274.30 due them according to the Treaty of 1855. In his report to the southern superintendent of Indian affairs, Rector made his case forcefully by describing the loss of crops and the intense suffering the tribes experienced the past summer. He then pleaded, "Humanity urges that the department should ascertain their condition and necessities, and that as we aided in sending food to starving Ireland, so we should preserve from destruction and misery these faithful allies."[35]

Peter Pitchlynn traveled to Washington, D.C., in 1860 as a member of the Net Proceeds delegation, a group that attempted to negotiate a reimbursement for property the Choctaws were forced to leave behind in their hasty removal from their Mississippi homeland. While in Washington, he received numerous letters about the suffering of the people, urging him and the committee to make haste in resolving the tribe's rights to its funds. One such missive from Choctaw chief George Harkins stated, "We have had a very severe winter—the stock are dying up by the whole sale—corn very scarce in parts of the Country—some of the Choctaw are now in a State of Starvation[. W]hat the poor Choctaw will do to live I can't see—Nothing to buy provisions with."[36] The appeals contain a note of urgent desperation, reflecting the Choctaws' growing impatience. One wonders how they managed to survive, especially with such a lack of communication and funding from the federal government. Tribesmen approached their leaders daily, inquiring as to when the commissioners would issue a decision on their claim.[37] Congress did approve $50,000 in relief to the tribes of Kansas and Nebraska in 1860, including the beleaguered Osages and Quapaws, as well as the Senecas and Shawnees, but the government was not yet prepared to allocate funds for the slave-owning tribes of southern Indian Territory.[38]

The absence of federal assistance placed an incredible burden on local resources. The Choctaw General Council did appropriate $134,512.50 for the purchase, shipment, and distribution of 65,000 bushels of corn to its people.[39] Still, many poorer families had to turn to their wealthier neighbors for foodstuffs.[40] It is no wonder that, in this situation, incidence of crime increased as people competed for the precious few resources available. As the rates of murder and violence rose, the courts became too full to prosecute offenses effectively, and the maintenance of law and order suffered as a result.[41] These factors influenced many to leave the region. During 1860 the crops looked promising in the spring, but harvest time brought a depressing loss. The Native people suffered intensely, and agents appealed to the federal government to ship in supplies to avert a complete breakdown of tribal society.[42]

With increasing concerns over law enforcement and the darkening horizon of internecine strife, it is not surprising that in Indian Territory, many land owners sold the improvements on their farms at reduced rates, allowing others to acquire substantial property rather cheaply.[43] In 1861 Elias C. Boudinot, a Cherokee tribal member, wrote to his uncle Stand Watie, asking his help in "purchasing improvements on grand River where large farms may be hereafter made. And also intend to purchase on the

Arkansas."[44] This flurry of property purchases was probably instigated by drought-reduced land improvement prices. As farmers sold their cattle and ate their supplies of wheat and corn seed, they limited their agricultural options, and they were thus willing to sell their improvements for subsistence money. Boudinot requested his uncle's aid because "a few hundred dollars with stock may be expended . . . which will in a short time return an immense percentage."[45] The stock in reference was most probably cattle, which would have indeed brought a handsome return on their investment: the region was largely devoid of them, and as moveable wealth, the bovines could be relocated to water rather easily.

The looming Civil War added considerably to the Indian Nations' hardships. In 1861, Union authorities recalled the troops from the frontier outposts of Fort Washita, Fort Arbuckle, and Fort Cobb. Under the new Republican administration of Abraham Lincoln, the federal government placed Indian Territory low on its list of priorities. Certainly, there were more pressing matters occupying the federal government's energies, but the Native people were desperate for relief, and none was forthcoming.

The Choctaws pose an important case study in the degenerating relationship between the U.S. government and the Five Tribes. Most of the Choctaws' trust funds were invested in government bonds, so they had no access to the money when they needed it most. They had to rely on the federal government to dispense monies to them. Shortly after Lincoln took the oath of office in March 1861, Congress approved funding for the Choctaws in the amount of $250,000, a portion of a larger award the government owed the tribe for damages incurred during their removal. The Choctaw General Council agreed that the greatest part of this sum would be used as relief for the starving members of their tribe. Unfortunately, hardly any of the money actually made it to Indian Territory. As hostilities erupted due to the secession of the southern states, the money was either lost or stolen on its way to Choctaw Nation.[46] The tribes of Kansas had received aid, but in Indian Territory only the Creeks and Chickasaws received any promise of money or relief, despite the fact that the government owed money to the all the tribes it had signed treaties with. This failure to appropriate funds for the destitute tribes of Indian Territory is probably due largely to the Interior Department's using the trust fund's government bonds for other purposes.[47] Perhaps the Union assumed that the slaveholding Five Nations would automatically swing to the Confederacy and wished to withhold any money that might make its way into secessionist hands.

Given this lack of federal action to relieve the effects of the drought, it seems that the Five Tribes would automatically side with the Confederacy. Culturally, the slaveholding tribes had more in common with the South. Plantations, slavery, and close geographical proximity would have made an alliance with the Confederacy seem a natural decision. Yet the tribes did not immediately vote to side with the Confederacy, mainly because such an act would destroy any hope of obtaining their treaty monies from the U.S. government. These federal funds were also influential in neighboring Arkansas. While serving as state commissioner, David Hubbard wrote that the state's western counties were reluctant to vote for secession as long as the Five Tribes were "still under the Government at Washington, from which they receive such large stipends and annuities."[48] Many residents of these counties no doubt had contracts with the tribes that were paid for by the government stipends. The neighboring state of Texas was also intensely concerned about the allegiance of Indian Territory in the impending national crisis.

The issue of Union or Confederate allegiance had yet to be settled among the Five Tribes. Not all members of the nations leaned toward supporting the seceding states. During the mid-1850s, tribal leaders of the Choctaw Nation tried to form a state and apply for admittance into the United States. The question that had to be addressed was what kind of state they would seek entrance as: slave or free? In 1857 Choctaw leaders wrote a tribal constitution that mandated humane treatment for slaves and gave the general council the power to control, or even prohibit, the movement of slaves into Choctaw territory.[49] A moderate tribal faction had obtained enough power to enter these clauses into the new constitution, revealing a divergence from their southern heritage. By 1860, the conservative faction of Choctaws had come to power and introduced a new constitution that omitted the provision for the fair and humane treatment of slaves and withdrew the council's power to regulate the movement of slaves into Choctaw lands.[50] This power shift, of course, pushed the Choctaw Nation closer to its eastern and southern neighbors and ruined any chance of obtaining statehood.[51]

Still, there was a strong tribal lobby for remaining in the Union. Peter Pitchlynn, after assuring the commissioner of Indian affairs that the Choctaws would remain neutral if war broke out, returned to Indian Territory and actively sought to convince his tribesmen of just such a course. But secessionists worked hard to muzzle the Union loyalists. The Texas Vigilance Committee coerced antisecessionist missionaries like John Edwards and

George Ainslee to flee Choctaw Nation and silenced Peter Pitchlynn with death threats for his efforts to keep the Choctaws neutral. George Hudson, another principal chief who advocated neutrality, was intimidated by a political rival's ardent speech attacking all who opposed secession. Although Hudson had prepared a message advocating neutrality, when he spoke, he called for the appointment of commissioners to meet with Confederate officials.[52]

Despite the secessionists' tactics, the Choctaw representatives refused the first Confederate overtures for an alliance. A Texas treaty commission composed of James Harrison, James Bourland, and Charles Hamilton traveled to Indian Territory to exhort the Five Tribes to join the Confederate cause. They began their mission on February 27, 1861, just eleven days after Twiggs surrendered the U.S. military posts in Texas. On March 10 the delegation arrived at Boggy Depot in time for a critical convention of the Choctaws and Chickasaws. The council convened on the next day and requested to hear from the Texas delegates. The Confederate spokesman, James Harrison, addressed the Choctaw Council, claiming that the U.S. government "had ceased to protect us or regard our rights."[53] The Choctaw delegates received Harrison's speech with some embarrassment, for they still had representatives in Washington attempting to withdraw the tribe's invested funds to more accessible local banks in case war erupted between the regions. At the end of the meeting, the Choctaw Council took no action. They were first interested in finding the results of the Washington delegation, as the drought had made it imperative to obtain funding, food, and seed for the next planting. The Confederacy would have to wait.

The Texas delegation hurried north upon receiving information that the Creeks and Cherokees were in similar meetings. They reached Creek Nation too late to take part in the convention, but Muscogee leaders did inform them of a proposed meeting of the Five Tribes scheduled for April 8 at North Fork, Creek Nation.[54] In the meantime, the Texan delegates thought they might as well visit the Cherokee leaders to ascertain their attitude toward the Confederacy. The delegates traveled to Park Hill and called upon Cherokee chief John Ross, but Ross was obviously waiting to see which direction the proverbial wind blew before he committed to any alliance. He treated the Texas men cordially, but persistently claimed the Union was not dissolved. The delegates reminded Ross of Secretary of State Seward's speech calling for the opening of Indian Territory to settlement and won a small concession for their efforts. Ross intimated that if Virginia and other border states joined the Confederacy, he would urge his tribe to do likewise.[55]

The Texas commission then traveled to North Fork for the anticipated Council of the Five Nations. But after month upon month of drought that dried and compacted the topsoil, making it mostly impervious to water, heavy rains made travel difficult as water ran off into streams and rivers, flooding the land. In these conditions the Canadian River became impassable, and the Choctaw and Chickasaw leaders were unable to cross and attend the meeting of the Five Tribes, but representatives of the Creek, Cherokee, Seminole, Quapaw, and Sauk tribes were at the meeting. James Harrison gave an impassioned two-hour speech that received little response. For all their work, the Texas delegation received no promises of alliances from any of the Five Tribes as they headed back to Texas. On their way, they noted the concentration of U.S. forces at Fort Washita in Indian Territory, a mere twenty-four miles from Texas, and in their report to Governor Edward Clark they claimed two to three hundred good men could take the post.[56] In fact, the position of Major William Emory, commander of the federal troops in Indian Territory, was tenuous at best. What was he to do if threatened with attack by Texas militia? Would the impending national conflagration begin in Indian Territory due to some act of poor judgment?

The standoff between the U.S. government and the Confederate authorities ended on April 12, when P. T. Beauregard ordered his South Carolina batteries to open fire on Fort Sumter at 4:30 in the morning. The surrender of the fort the following day forced some waffling states to make clear decisions. Virginia quickly cast its lot with the Confederacy on April 17, and Arkansas opted to secede on May 6, followed by Tennessee and North Carolina. With both Texas and Arkansas in the Confederate fold, pressure mounted for the Confederate States to open negotiations with the tribes of Indian Territory. Confederate Secretary of State Robert Toombs tapped Albert Pike for the position of special agent to the Five Tribes. Pike, already familiar with and trusted by the people of Indian Territory, was a good choice for this office. After stints as a teacher, fur trapper, and journalist in cities as different as Boston and Santa Fe, Pike had been admitted to the bar in Fort Smith and had represented the Creek and Choctaw tribes in their pursuit of federal payments, earning the Natives' trust.

Meanwhile, U.S. officials finally saw the danger of keeping troops in Indian Territory, and ordered Major Emory to withdraw the forces under

Albert Pike the Younger, 1865. Glass Negative. Courtesy of the Brady-Handy Collection, Prints and Photographs Division, Library of Congress.

his command to Kansas. By mid-May, he had successfully removed federal troops from all the posts in Indian Territory, and these were quickly occupied by Confederate forces. This abandonment of the region's forts disheartened Native Unionists and gave Indian Territory secessionists another point to make about broken promises from the federal government. It was at this moment of Unionist weakness that the Confederacy sent their agent, Albert Pike, to negotiate treaties with the Five Tribes.

Pike began his tour of the Five Tribes by calling on John Ross at Park Hill, as gaining Cherokee allegiance was key to Confederate strategy. Cherokee land bordered northwestern Arkansas and the Union state of Kansas. Securing Arkansas's western boundary would allow the state some breathing room. Further, the Cherokees were the most divided tribe concerning remaining in the Union or allying with the Confederacy. The old tribal divisions were reopened with the events of 1860–1861. The Treaty Party activists, Stand Watie, Cornelius Boudinot, and others sympathized with the southern states, but the man in power, Chief John Ross, was more determined than ever to wait and see how events unfolded. Historians often describe this division as if all full-bloods followed Ross and all mixed-bloods supported the Treaty Party, but in actuality it was not that simple. John Ross, as his name might imply, was only one-eighth Cherokee and owned slaves, yet he represented the interests of the full-bloods and leaned toward supporting the Union. The contradiction is glaring. There were certainly at least two power groups seeking to control the politics of Cherokee Nation. Those in favor of maintaining the treaties with the United States began wearing pins in their coat lapels to identify their political position to others, winning the moniker of "Pins" for this act of solidarity. Pike claimed that Ross had 1,000 to 1,500 followers, exceeding the Texas Delegation's claim that only 20 percent of the tribe supported Ross in his adherence to a policy of neutrality.[57]

After his failed attempt to convince Ross to side with the Confederacy, Pike traveled across the Arkansas River to the Creek agency. Although he did not expect any difficulties in negotiating with the Muscogees, in truth the tribe was every bit as divided as the Cherokees.[58] Old animosities remained, dating back to the assassination or execution, whichever way one views it, of Lower Creek chief William McIntosh after the signing of the Treaty of Indian Springs. Creek chief Opothleyaholo, who had warned McIntosh that signing the treaty would be equivalent to signing his own death warrant, participated in the murder of McIntosh and then made the trek to Indian Territory. Opothleyaholo's stature had grown among the Upper Creek, while the Lower Creek, led by Moty Kinnaird, still bore a grudge for the acts committed that fateful day in 1825.

Fortunately for Pike, when he arrived in Creek lands, Opothleyaholo was in western Indian Territory meeting with the Plains Tribes at Antelope Hills. The chief was trying to form a confederation of Indian tribes that would agree to remaining neutral through the coming internecine conflict.

With Opothleyaholo absent, neutrality lost its best spokesperson. The first chief of the Upper Creeks was Echo Harjo, a slaveholder, and both he and Moty Kinnaird were frustrated with the U.S. government for withholding treaty payments. Given these circumstances, it is not surprising that Pike successfully negotiated a treaty of alliance between the Creeks and the Confederate States on July 10, 1861, at North Fork village.

Pike had also sent communications to the Choctaws and Chickasaws to meet him at North Fork and hoped in vain to hear from Cherokee Confederate sympathizers, but Treaty Party members of the Cherokee Nation feared they would pay with their lives if they met with Pike.[59] The Choctaw Council on the other hand had received the Net Proceeds delegates back from Washington, unsuccessful in their mission to transfer tribal funds to southern banks. Two days after the Creeks signed their treaty with Pike, delegates from the Choctaw and Chickasaw tribes signed a treaty of alliance with the Confederacy as well, tying their fate to the Confederate States of America. With these negotiations terminated successfully, Pike made his way to the Seminole Agency. The Seminoles had strong ties to the Creek Indians, and though tolerant of slaveholding, most Seminoles supported neutrality.[60] Pike made sure that influential southern sympathizers were at the council to help convince the Seminoles to ally with the South. Moty Kinnaird and Chilly McIntosh, the oldest son of executed Creek chief William McIntosh, were present, as were Seminole Indian agent and southern sympathizer Samuel Rutherford and a friendly Indian merchant from Fort Smith, Charles B. Johnson, who also had pro-Confederacy leanings. Still, it was not until August 1, 1861, that Seminole chief John Jumper agreed to a treaty with the Confederate States.[61] Shortly after the signing of the Seminole-Confederacy treaty, Opothleyaholo returned from the Antelope Hills, and pro-secessionists feared that he would organize resistance to the treaty.

Unflappable, Pike proceeded to Fort Cobb to treat with the Leased District tribes who had been so roughly treated by Texans just two years earlier when living on the Brazos Reserve. These negotiations differed from previous efforts in that all Pike requested of these tribes was neutrality. Some bands of Comanches showed up at the talks, so Pike ensured they would restrain from raiding into Texas in exchange for rations and gifts.[62] In the meantime, on August 10 the Confederacy defeated Union forces at the Battle of Wilson's Creek in Missouri, sending them reeling back towards Saint Louis and bolstering Pike's negotiation efforts.

On his return to Fort Arbuckle, Pike received a message from Chief Ross informing him that the Cherokees were holding a council on August 21 and that Ross was ready to sign a treaty. The accords between the Confederacy and the Creeks, Seminoles, Choctaws, and Chickasaws had isolated the Cherokees, and the withdrawal of U.S. troops from Indian Territory added to the recent defeat of federal forces at Wilson's Creek had left the Cherokee vulnerable if they remained neutral. The headman had little choice but to negotiate with the Confederacy. Pike and Ross finalized the treaty on October 7, 1861. These agreements between the Five Tribes and the Confederate States of America, though differing in some individual terms, bound all the Five Tribes to the Confederacy and required the tribes to furnish troops but guaranteed that no Indian companies would be required to serve outside Indian Territory. The Confederacy also agreed to take on the financial burden of arming the Indian musters. In fact, Pike had exceeded his authority in negotiations. A pre-treaty communication from the Confederacy's secretary of war to Pike made clear that he was not to incur any financial burden on the Confederate States through the treaties, but he did just that when he promised that the Confederacy would honor the tribes' treaty payments and furnish the Native recruits with arms.[63]

The Confederacy had worked hard to negotiate treaties with the Five Nations, and with good reason. The Indian Territory was strategically located in that it flanked the state of Arkansas and buffered Texas from any Union invasion from the north. Indian Territory could also be the jumping-off point for a southern invasion of Kansas. Trails through Indian Territory shortened the time it took to get Texas cattle, horses, and other goods to Tennessee.

In conclusion, Native people had many reasons to distrust the U.S. government. Their forced removal by government troops combined with the failure of the federal authorities to meet their treaty obligations and to protect the tribes from outside aggression certainly qualify as strong motives for Natives to abandon any loyalty to the federal government. The Five Tribes also held many ties to the South. Geographic proximity dictated that the tribes consider invasions from Texas and Arkansas if they sided with the North, and most of the river-borne commerce from Indian Territory was destined for southern markets. Many tribal members had secessionist relatives who lived in southern states, and the U.S. Indian agents, who were for the most part staunch southerners, may have had a certain degree of influence among the rank-and-file tribal citizens as well. There was also the "peculiar institution" of slavery that wealthier members of the Five Tribes practiced, but this system of coerced labor was not adopted by the majority of tribal members. Although the elite did hold an inordinate amount of influence over tribal politics, it was the common folk who would form the bulk of the troop levies fighting for the Confederacy.[64]

There were other reasons the Five Tribes distrusted the federal government and allied instead with the Confederacy. In 1860 Secretary of State William Seward gave a speech in Chicago titled "The National Idea, Its Perils and Triumphs," arguing that "the Indian Territory, also, south of Kansas, must be vacated by the Indians."[65] Many Natives of Indian Territory felt this statement was representative of Lincoln's agenda.[66] Still, underlying the political and social considerations was the stark fact that seven years of drought had brought the Indian Nations to a desperate point. They were in dire need of relief, and the federal government refused or poorly managed its promises of funds for the tribes. So when Confederate emissary Albert Pike visited them promising money, sovereignty, and political participation, the tribes reluctantly agreed to support the Confederacy.[67]

Pike well understood the primary motivation behind the Five Tribes' alliance with the Confederacy. He wrote to his superiors that in order to obtain the tribes' support, the Confederacy would need to "guarantee them their lands, annuities, and other rights under treaties" they had already made with the United States.[68] Jefferson Davis also understood the significance of annuities: in a speech to the Confederate States Congress, President Davis stated that the pecuniary obligations of these treaties were of great importance.[69] Davis and Pike convinced the Confederate Congress to promise the Choctaw Nation a perpetual annuity of $9,000 and assure the Chickasaw a yearly payment of $3,000, which, although substantially less than the $265,927.55 the U.S. government owed them, was something.[70]

Thus the Five Tribes signed the Faustian bargain that sealed their fate to that of the Confederate States, to the surprise of many in the North. The New York Times ran an article in July of 1861 that belittled the importance of the Confederate treaties with the Five Nations of Indian Territory and cavalierly mentioned the privation of the tribes: "The Indians have no warriors to spare to conduct hostilities against the Lincoln government; and how are they to be saved from starvation is a problem that now concerns only the Southern Confederacy, whose standard they have joined."[71]

The Times article cited the Galveston News in claiming that the Chickasaws had voted unanimously to secede from the United States. Yet, six months later, in December, a delegation of Creeks, Seminoles, and Chicka-saws traveled to Washington, D.C., to see if their treaties with the United States, with their attendant annuity and other monetary concerns, were still valid. The delegation claimed that a small minority of their tribesmen, one hundred total, had voted to withdraw from the Union, but that 1,240 had voted to remain.[72] In the absence of voting records, one is left wonder-ing who told the truth. Both the Confederate sympathizers in Galveston and the Washington delegation had reasons to skew the numbers. Still, the voice of the Indian delegation seems the more reliable. In the case of the Choctaws, the influence of the slave-owning minority, only four percent of the whole tribe, was disproportionately strong.[73] Missionary Cyrus Kings-bury certainly believed the slaveholding Indians overshadowed the slave-less majority in negotiations. He describes Albert Pike's inordinate political clout and the Choctaws' trust in the Arkansas attorney in a letter to fel-low missionary Reverend Treat: "We regret the present as an eventful crisis with the Choctaw. I have informed you how it happened, that the pres-ent General Council were elected by a small minority of the nation. They are mostly half breeds and are very much under the influence of Arkansas lawyers."[74] The reference to Albert Pike does not go unnoticed. There were not many options open to the tribes of Indian Territory in 1861: they could remain loyal to the Union and starve for another year, or they could sign an agreement with the Confederacy and hopefully relieve the suffering of their people. They chose the latter and bound their future to that of a tenuous Confederacy, all for bread.

EIGHT

The Early Civil War Years in Indian Territory

War is never a neat and tidy affair. There are no "smart wars" in which only military participants suffer injury. Instead, war unleashes a maelstrom of agony upon willing participants as well as countless innocents. The Civil War was no different, regardless of how honorable the cause was according to either side; it was war. Although every theater had its episodes of wanton violence, the Civil War in Indian Territory had a reputation for being particularly barbarous.[1] Some scholars attribute the viciousness to the participation of the "savage" Indians who were unable to curb their bloodlust when the war commenced, but the savagery of the war in Indian Territory was by no means confined to its Native participants. For others the cause was Indian Territory's close proximity to Bleeding Kansas, with the violence of that sad chapter of history spilling across the border into Native lands; it is well known that William Quantrill and other paramilitary groups from the Kansas conflict crossed through Indian Territory on their way to Texas to escape federal forces.[2] Another reason for the incredible violence could be the strained relations within tribes and between Native power groups that predated the war, an explanation particularly viable in the case of the Cherokee and Creek nations, whose pre-removal history was rife with internal divisions. There were also occasions when old Native rivalries brought out the worst in the participants, as when Stand Watie ordered the torching of Cherokee Chief John Ross's home, but the devastating drought also played its role. Competition for resources can turn nasty when self-preservation is at stake. The extreme aridity influenced military strategies, inhibiting successful invasions from the north and reducing much of the later efforts of the war to cattle raids.

After the withdrawal of U.S. troops from their posts in Indian Territory, the Texas militia quickly occupied Fort Washita, and Arkansas troops took Fort Smith. These actions placed the Native people who wished to remain within the United States in a precarious position. Indeed, even those who espoused strict noninvolvement felt threatened. Opothleyaholo, the Creek champion for neutrality throughout the war, collected his followers, their possessions, and livestock and began moving to a more secure location closer to the Kansas border. It appears that the Muscogee leader, by then eighty years old, had plans to construct a fort on Walnut Creek and wait out the war.[3]

For his success in negotiating the treaties with the tribes of Indian Territory, Confederate authorities placed Albert Pike in command of the district, but at that time he was away visiting Richmond, trying to convince the provisional government to fund his treaties. In Pike's absence, he placed Douglass Cooper in temporary charge of Indian Territory for the Confederacy. Cooper had served as agent to the Choctaws and Chickasaws prior to the war and now could add to his laurels a commission as colonel in the Confederate army, commanding the Choctaw and Chickasaw Mounted Rifles. Pike had requested white troops to bolster the Indian units, and Texas provided three regiments under the command of General Ben McCulloch who, respecting the wishes of the Chickasaw Nation, withdrew most of his Texas troops from Fort Washita and took his command to Little Rock, Arkansas, where he set up his headquarters and endeavored to organize an Army of the West.[4]

In the meantime, Opothleyaholo had been in contact with Union officials in Kansas, stressing his need for protection from Confederate forces.[5] In addition to his own followers, he had accumulated some Seminole families, who traveled with all the possessions they could carry and their livestock. To defend this exodus of refugees he had at most 2,500 warriors. The presence of a large pro-Union population was a reason for alarm to Douglass Cooper, who moved to intercept the Unionist Creeks with a large portion of his command, some 1,400 troops. Pike preferred not to use Indian units to fight other Indians, but he felt there was no alternative with Mc-Culloch's forces stationed almost three hundred miles away in Little Rock.[6] Cooper claimed he had attempted to negotiate with the Opothleyaholo but was rebuffed at every effort.[7] Still, Cooper was determined to either convince Opothleyaholo to submit to the terms of the treaty or drive him out of Indian Territory. The Creek chief's refusal to honor the Confederate treaty signaled the beginning of hostilities; setting the tone for the entire war, this first campaign targeted civilians.[8]

November 19, 1861, was the first official bloodletting of the Civil War in Indian Territory: Cooper's scouts located Opothleyaholo's position in the northern limits of the Creek Nation, and Confederate troops charged the camp but found it deserted, though some of the men sighted a few neutral Creek scouts and gave chase as they fled. About four miles from the Creek camp, the scouts entered a timbered area. As the Texans rode up to the woods, gunfire erupted, and they found to their surprise that they were greatly outnumbered by neutral Indians. Six men in Cooper's command perished in the skirmish; the number of casualties for Opothleyaholo's followers could not be substantiated, though Cooper claimed the loyalists suffered 110 killed and wounded.[9] The Creek leader's men successfully defended the retreat of their women and children. However, they vacated the camp so hurriedly that they left behind "the chief's buggy, 12 wagons, flour, sugar, coffee, salt &c., besides many cattle and ponies." With winter fast approaching, the neutral Creeks and Seminoles would sorely miss these vital supplies.[10] This brief engagement at Round Mountain near present-day Yale, Oklahoma, forced the neutral Indians to relocate, but, resolute to stay in their new homeland, they did not yet flee to Kansas. Opothleyaholo obviously thought he could adequately defend his people while remaining in Indian Territory.

Cooper chose not to pursue Opothleyaholo. The neutral Indians' livestock had consumed the paltry drought-weakened grasses in the vicinity of the camp and the Confederate horses were spent, so his first priority was to find sufficient pasture on which to feed his mounts. At this point, Cooper received a dispatch from McCulloch ordering the Indian Regiment to take a position near the Arkansas line, since the Union commander in Missouri, John C. Frémont, was at Springfield with a large force and expecting to move south. With this Union threat, McCulloch might need Cooper's forces nearby as reinforcements.[11] However, Frémont instead retreated towards Rolla, Missouri, freeing up Cooper to continue pursuing

Opothleyaholo. Accordingly, the Confederate colonel marched 780 of his men to Tulsey Town, modern day Tulsa, on the Arkansas River.

A bedraggled prisoner escaped from the neutral Creek camp informed the Confederates that Opothleyaholo was planning to attack with two thousand warriors. As can be imagined, such figures were consistently exaggerated during the war by both sides, but in any event, the threat of attack remained serious. Preparing his defense, Cooper ordered the Cherokee Mounted Rifles under Colonel John Drew to meet up with his command; however, the pro-Confederate leanings of Drew and his men were tenuous at best. Cooper felt the threat of Opothleyaholo's attack real enough to gamble on the loyalty of Drew's command. The two Confederate units met on December 8, 1861, and camped two miles apart on Bird's Creek, north of Tulsa. Drew informed Cooper that Opothleyaholo had sent a messenger asking for a peace conference, placing Cooper in a predicament. He was worried that the neutral Indians could muster some two thousand men, greatly outnumbering his force, and that if pushed too hard, Drew's unit might desert or, worse, defect. After assuring Drew that he did not want to shed more blood, Cooper ordered a party of four men, Major Pegg, Captain George Scraper, J. P. Davis, and Reverend Lewis Downing of Drew's Pin Indians, to travel to the neutral Indian camp and set up a time and location for negotiations. That evening, the majority of Drew's command deserted. According to the story told to Cooper, the Cherokee envoy had returned without being able to speak with Opothleyaholo because he encountered "several thousand" "painted warriors" preparing to surround Drew's camp. When the messenger ran back to the Union Cherokee camp, telling everyone of the impending attack, the Pin unit melted into the night, leaving behind tents, horses, and even, in some cases, guns.

Rumor can certainly strike fear into any military camp, but this episode smacks of collusion. Opothleyaholo was a wily veteran who might have used sympathizers within Drew's command, if not Drew himself, to spread wild rumors and greatly inflate the neutral Indian numbers. That the Cherokee Mounted Rifles left behind horses, tents, and especially guns, defies logic. Even if the enemy was at the gates, so to speak, one would expect the Cherokees to grab their arms to defend themselves, but abandoning all these essentials would allow the needy Creek and Seminole refugees to acquire them later. Although Drew and twenty-eight men visited Cooper, claiming they would remain true to their treaty and assist in defending the Confederate camp, the Confederate colonel acted quickly to have the

Cherokee supply wagons brought to his camp.[12] The neutral Creeks insti-gated no combat that night, Major Pegg had made his way to Fort Gib-son following the mass desertions, and George Scraper and his company deserted, indicating, according to Drew, that they "doubtless were in the camp of Hopoeithleyohola."[13]

The next day, in order to maintain open communication with reinforce-ments that were on the move to the region, Cooper broke camp at Bird Creek and relocated closer to Tulsey Town. As his troops were moving south along the creek, they stumbled into a trap laid by Opothleyaholo at Chusto-Talasah, or "High Shoals," where the Muscogee leader had placed his warriors along a high bank of Bird Creek. In some places thirty feet tall, affording only a few places to access the top, this natural feature made an imposing defense. Opothleyaholo then sent a group of warriors to attack the Confederate forces and retreat, leading them into an ideal location for a defensive battle. It took hours to dislodge Opothleyaholo's men, giving their women, children, and livestock precious time to move northwest towards Kansas. The battle raged for four hours, interrupted only by the setting sun. Cooper's pursuit of the fleeing refugees was hindered by a shortage of ammo, and as night fell across the darkling plain, he ordered a withdrawal.[14]

As Opothleyaholo and his people moved northwest, Cooper marched southeast to Fort Gibson and its store of supplies and ammunition. The old Muscogee leader had proven a tough nut to crack, and Cooper decided he needed help. He sent a dispatch to Little Rock asking General McCulloch for reinforcements, but the renowned Indian fighter was away from head-quarters. So Cooper appealed to the next person in the chain of command, Colonel James McIntosh, who commanded McCulloch's division in Little Rock.[15] Colonel McIntosh led these fifteen companies of Texas and Arkan-sas soldiers, about two thousand armed men, on a forced march to Fort Gibson, and there conferred with Colonel Cooper. The two agreed to move north in two wings: Cooper would travel up the Arkansas to cut off the neutral Indians' retreat, while McIntosh marched north along the Verdigris to locate the enemy camp. Due to the persistent drought and winter season, forage was extremely scarce, and both commanders agreed that either con-tingent should attack immediately and not wait for the other.[16]

In 1861 Christmas day closed on the frontier of Indian Territory without the pleasant sound of peaceful noels, but instead with the threat of more violence. McIntosh's 1,380 men prepared to make camp after a long ride, but the site of neutral Indians in the distance sent the Confederate soldiers scurrying for their arms. McIntosh sent a regiment out, but when the Natives turned and ran, he knew it was a ruse. Just as the killdeer will flap in distress as if she has a broken wing, leading would-be predators away from her nest, the neutral Creeks were attempting to lead the Confederates on a wild goose chase, away from the Native camp.

Undeterred, the next morning McIntosh ordered his men to continue to the location their intelligence had identified as the site of the Creek camp. But overnight, another mishap plagued Cooper: his teamsters had deserted, making it difficult for him to rush to McIntosh's aid if needed. Close to noon, McIntosh's troops crossed Hominy Creek and were greeted by the sounds of musket blasts, puffs of smoke, and the smell of gunpowder. Opothleyaholo's warriors had taken a strong position on a rugged hill overlooking the ford and enjoyed the protection of its steep wooded slope. The Confederates charged, engaging for a time in fierce hand-to-hand combat before they finally reached the summit of the knob. At this point, Opothleyaholo's warriors broke and ran for their camp to make another stand, but a cavalry charge by the Confederates dispersed the remnants of the neutral Indian defenders.

By 4:00 in the afternoon, McIntosh found himself in control of Opothleyaholo's camp. Here he found some letters addressed to Opothleyaholo from E. H. Carruth, agent to the Creek and Seminole tribes for the U.S. government. These letters promised refuge in Kansas, but what worried McIntosh was the possibility that Opothleyaholo's warriors would participate in Union invasions of Indian Territory. McIntosh used these letters to justify his attacks on civilian men, women, and children. In his report, he claimed Confederate losses were eight killed and thirty-two wounded, while he placed the enemy dead at 250.[17] Being in control of the field of battle, McIntosh should have had an accurate casualty count, but these one-sided numbers are difficult to believe. In any event, it is certain that Opothleyaholo's warriors were desperate, once again, to cover the retreat of their women and children, and thus paid for the defense of their camp with their lives. This engagement, known as the Battle of Chustenahlah, broke the neutral Indians. They lost their tents, wagons, livestock, and food and had to flee north to Kansas on foot as the weather turned cold and snowy. Stand Watie and his three hundred–strong Cherokee unit arrived too late to participate in the battle, but he tried to make up for it by vigorously pursuing the refugees the next day. Cooper, likewise, rounded up 150 straggling refugees, mostly women and children, while "the weather was exceedingly cold; sleet fell in considerable quantities during the day, and there [was] the appearance of a snow-storm [in the offing]." The conditions were so frigid that one of his men froze to death while his troops scouted the region north and west of Tulsey Town.[18] The suffering of the defeated neutral Indians, who had lost loved ones and everything else, must have been even worse; their shelter, horses, and food had been confiscated by the Confederates, and as they trudged toward Kansas, the weather brought ever more snow and frigid temperatures. One is hard-pressed to find armies targeting non-combatants in other theaters of the Civil War, supporting the interpretation that the conflict was even more savage in Indian Territory than elsewhere. This can be attributed to several factors: the presence of internecine conflict within the tribes; the absence of institutions of law and order, resulting from tribal displacement during the war; and the fierce competition for resources in the confines of a prolonged and severe drought.

The story of these refugees is filled with adversity, caused for the most part by poor Union administration. During their first winter in Kansas, the neutral Creeks and Seminoles suffered even more deaths due to lack of shelter and food, though they were purportedly under the care of the U.S. government.[19] It took a month for Congress to approve of funding for the relief of the Creek refugees. In that time 240 Muscogee Creeks died, and camp surgeons amputated more than a hundred frostbitten limbs.[20]

A change in command of Union forces in Missouri replaced Frémont with Samuel Curtis, a West Point graduate and veteran of the Mexican-American War. Curtis moved his troops straight toward Springfield, Missouri, and the commander of Confederate units in that border state, Sterling Price, was in no condition to put up a fight. Price and McCulloch had feuded over how best to use their combined forces, and their incessant bickering eventually led the Confederate War Department to appoint Van Dorn as the officer in charge of the Trans-Mississippi Theater. This inability to coordinate efforts led Price to retreat south from Missouri, while Curtis marched his men all the way into northwestern Arkansas, near Pea Ridge, without opposition. Now Union troops threatened Fayetteville.

Meanwhile, Pike had left Richmond with his treaties ratified, with promises of Enfield rifles and, after a side trip to Charleston, South Carolina, with $95,000 in gold to disburse to the allied tribes, he finally returned home to Little Rock in late January 1862.[21] He met with Major General Earl Van Dorn who, as commander of the Trans-Mississippi Department, had replaced General McCulloch as Pike's immediate superior. Pike claimed that at this meeting Van Dorn had promised him "sole control of Indian country" and three regiments of Arkansas infantry.[22] That promise became a misunderstanding that eventually led to Pike's resignation.

When Van Dorn assumed command, he decided to take direct control of both Price's and McCulloch's units and began organizing the defenses. He ordered Pike to rush all the units under his command to Pea Ridge, Arkansas, which put Pike in a difficult position, as it abrogated the treaties he had made with the tribes promising them that their men would not be asked to fight outside Indian Territory. In February of 1862 Pike met with the Indian troops mustering for service in the Confederate army at Cantonment Davis. All of the men were desperate for money to purchase food for their deprived families. The drought had forced most prices up and family gardens had not produced as much as was needed. As a result of the financial difficulties, many Indian families had gone into debt to local store merchants.[23] Such problems made matters more difficult for Pike. The Confederate Congress had given him money earmarked for certain tribes, but dispersing it to the desperate Indian soldiers would leave him unable to use the funds as he saw fit, namely, to purchase weapons or provisions. Compounding Pike's worries about the tribal monies, a courier arrived with Van Dorn's orders to march his troops to Bentonville, Arkansas, a directive sure to be unpopular with the Indian troops, who had been promised they could fight from Indian Territory.

Pike delivered the earmarked funds to the Comanche, Osage, and Reserve Indians gathered at Fort Davis. Aware of Pike's disbursement of annuities, groups of enlisted Chickasaw, Choctaw, and Creek men traveled to Fort Davis in the hope that they would receive their pay as well. When these men demanded their promised money, Pike had to explain to them that the Confederate legislature had designated the money for specific tribes. The frustrated Indian soldiers began to pack up and leave upon hearing Pike's explanation, refusing to march for Bentonville without their pay. The dire situation demanded that Pike break protocol. He paid the Choctaws and Chickasaws and promised the Creeks that they would receive their money soon. The disbursements delayed Pike's march to Arkansas for three days.[24]

In Arkansas the military situation worsened for the Confederacy. General Curtis's Union troops, bolstered by reinforcements, chased the Confederate units under General Price out of southern Missouri into northwestern Arkansas and threatened to advance on Fort Smith. Van Dorn rode to Pea Ridge to take command of the Confederate forces personally, but when crossing the Little Red River, his canoe tipped over. He emerged from the river drenched, but seemingly none the worse for wear. However, a hard ride across the Ouachita Mountains in frigid temperatures took its toll on his body. As he arrived in Van Buren, he fell ill with a fever, and after continuing his journey to the Confederate position near Bentonville, he became so weak that he could not even mount his horse.[25] Still, Van Dorn was determined to defend the northern approaches to the Arkansas River, the main artery of supplies to Forts Smith and Gibson, but he would have to deal with Union troops who occupied a strong position in the area. Curtis had set up the majority of his units along a bluff on the northern bank of Sugar Creek, making their position too strong to take by a head-on assault.

Pike and his regiments arrived at the Confederate camp on March 6 without the lagging Choctaw and Chickasaw units, and Van Dorn quickly attached these Indian troops to McCulloch's command. In a brilliant ruse, Van Dorn ordered McCulloch's men to leave their campfires burning the night of March 7, 1862, while they conducted a predawn march to Elkhorn Tavern to force the Union army's flank. However, the night march was delayed by a botched creek crossing. Thus, Van Dorn's men were not in position until 10:00 A.M., giving Curtis plenty of time to begin transferring Union units to defend his flank. In that day's fighting, Pike's unit, comprised of a thousand men, 80 percent of whom were Native, achieved one of the lone successes of the Confederate army when it charged and took a Union battery. Watie's and Drew's Cherokee regiments, along with the Texans, held their position for twenty confusing minutes as the inexperienced Indian troops' exuberance on taking an enemy position manifested itself in a celebration that rendered them deaf to any orders.[26]

Unable to consolidate his command, Pike ordered his men to fall back when Union artillery opened fire on them, and the Union infantry retook their position. The death of General McCulloch during a reconnaissance of Union positions prior to a counterattack left the Confederate troops without a cohesive command. Van Dorn could not rise from his sickbed, and Pike was not in communication with Confederate headquarters. As the Union army pressed its advantage, the Confederates retreated. Although the battle continued into the next day, it was a resounding victory for the North.

In the aftermath of the battle, Union officers began to report that some of their dead had been scalped. When the accusations reached Van Dorn, he replied that there must have been some mistake. The scalpings occurred at the site of the Indian regiments' twenty-minute celebration, but these were Cherokee, who were "civilized." In the investigation that followed, Union officers found that eight of the twenty-five soldiers killed at that specific location had been scalped.[27] Shocked and embarrassed by this situation, Van Dorn omitted the Indian troops' contribution to the Confederate effort in his official report to Richmond, another act that Pike felt was a slight to his men. Pike claimed he knew nothing of the scalping and that the only atrocity committed in the fight he was aware of was the murder of a wounded prisoner by a white bugler of the First Choctaw and Chickasaw. Pike ordered the arrest of the bugler, but when no witnesses stepped forward, the trial became a moot point, and later the accused perpetrator escaped.[28]

Wiley Britton, a private in the 6th Kansas Volunteer Cavalry and a native of Newton County, Missouri, the scene of savage guerrilla warfare throughout the war, traveled with the Union army to the Pea Ridge battle site in 1863. The northern military had incorporated neutral Indians into their service, and Britton, perhaps remembering the atrocities of that spring day, claimed the Union Indians committed fewer depredations "than the same number of white troops" while they were in Arkansas. In fact, according to Britton, it was a "very rare thing to hear of a complaint being made against our Indian soldiers for having committed unauthorized acts."[29] Such a statement, of course, does not absolve the Confederate Indians, but it does suggest that white soldiers could be as responsible for atrocities as Native American ones.

The Battle of Pea Ridge, as it came to be known, was the largest of its kind west of the Mississippi River. More than 25,000 men from both armies exchanged gunfire for two days, resulting in 2,400 casualties. The Union victory was extremely influential for Indian Territory. Following the military defeat, the Confederate high command shifted Van Dorn and most of his troops east across the Mississippi to bolster the Confederate Army of Tennessee; however, it simultaneously exposed the eastern flank of Indian Territory. The reassignment of Confederate troops was simply a matter of priorities, for though the Union army had enough troops to fully man all three theaters of the war, Virginia, the Tennessee Valley and the Trans-Mississippi West, the Confederacy did not. The shift of Van Dorn's units east meant that the Confederate high command deemed the Tennessee campaign a higher priority than the defense of Arkansas or Indian Territory. But to Pike, the relocation of Van Dorn's troops was another broken promise. The treaties he had signed with the tribes promised the Confederacy would provide munitions and troops for the defense of Indian lands. Now even Arkansas would have to defend itself alone.

As the drought intensified, Pike worried about the coming harvest. Most of the region's men were serving in the military, so there were few field hands to gather in the crops, vital to the last seed in a grain-starved area. In an effort to solve this dilemma, Pike granted furloughs to half of his fighting force, which consisted of a few non-Indian units from Texas and Arkansas, so those men could "return home to reap the wheat harvest." According to the general, this shortage of manpower and of corn made it difficult for mules to pull ordinance and supply wagons, complicating the resupply and reinforcement of Fort Smith and Arkansas.[30] The shortage also created the perfect conditions for price gouging, and Arkansas contractors began charging exorbitant rates for corn, beef, and other provisions. As a result, Pike skipped the middle men and organized his own system of purchasing supplies in Texas, sending a man from his unit to make the purchases and hire freighters to haul the supplies north.[31] Still, the scarcity was so acute that what little corn there was went to fulfill human demands, leaving the needs of livestock untended and further burdening the food supply, as the surplus cattle kept near Fort Smith for military consumption starved to death.[32] The Confederate response to this quandary was to let the cattle graze in the surrounding countryside, but that left the herds vulnerable to cattle rustling by outlaws and enemy soldiers.

Considering his supply lines to Texas untenable at Cantonment Davis, Pike withdrew his forces south to a location in Indian Territory forty miles from Sherman, Texas, across the Red River. Here he constructed Fort McCulloch, named after the general who had fallen at Pea Ridge. Although Pike's retreat left the Cherokee, Creek, and Seminole lands open to Union invasion, he was closer to the sources of provisions for his troops. Pike understood that his alliances with the tribes were resting on shaky ground. He worried that the Pin Cherokees would defect without a Confederate presence in their territory, and he was troubled that some of the ammunition, uniforms, and weapons he had purchased for the troops of Indian Territory had been intercepted and pilfered by other Confederate units in Arkansas.[33] Supplies were essential to keeping a poor and starving army in the field. In fact, such was the suffering from drought-induced famine that many Indian volunteers had signed up for the Confederate army solely in expectation of three square meals a day. Pike understood the situation well, stating on June 30 that "I am feeding with supplies of flour and bacon purchased in Texas most of the Indian tribes, and if I were to send back one train without provisions, and tell them I had no money and was not allowed to buy on credit, they would instantly disperse."[34]

Meanwhile, Van Dorn had chosen Major General Thomas Carmichael Hindman as Pike's replacement to command the Trans-Mississippi District, and he took command in late May. Realizing a sizeable Union force remained to his north, Van Dorn ordered Pike to send all of his non-Indian units to Little Rock, Arkansas. Pike was reluctant to let them go, but after writing his protests, he acceded to the order.[35] His following feud with Hindman would only heat up in the days afterward, corresponding to the worsening climactic conditions in Indian Territory.

Aridity has a lot to do with causing high temperatures. Humidity keeps temperatures down, but cloudless skies and dry conditions during the day allow the temperatures to rise above the century mark. It is not uncommon for the eastern portions of present day Oklahoma to reach over 100 degrees Fahrenheit during the early summer, but in late June of 1862 the readings at Fort McCulloch, taken at 1:00 P.M., registered 104, 105, 108, and 112 from June 27 through 30.[36] The drought's effects were now being felt even further east in Arkansas. Hindman claimed that due to the lack of corn as feed for horses, he had to "dismount four regiments of Texans and three of Arkansians." This decision was met with widespread disapproval and desertions. He also reported that many were deserting because they had "wives and children suffering for food" back home.[37]

Seizing an opportunity, Union forces prepared for an invasion of northern Indian Territory. The withdrawal of Confederate troops from the region disaffected half the Cherokee Nation. With Pike far to the south of the Canadian River, the timing seemed perfect for a Union success. On the other hand, Pike knew that his enemy's task would not be as easy as expected mainly because of the drought. There was "not an ear of corn to be had" in the environs of Fort Gibson, and Fort Smith only subsisted by the flow of provisions from north Texas.[38] Pike thought these conditions in themselves might present the best defense of Indian Territory: "I hope that the excessive drought, the utter destruction of corn and grass, the intense heat, and the scarcity of water may prove our best allies. No large force of the enemy can march now any distance into this country."[39]

On June 28, 1862, the anticipated invasion began. Some 6,000 Union troops, known collectively as the Indian Expedition and including units of loyal Indians commanded by Colonel William Weer, entered Cherokee land from Baxter Springs, Kansas. Brigadier General James Blunt, commanding the Department of Kansas, hoped to use the broken treaty promises of the Confederacy to win over rank-and-file Cherokees. He ordered Weer to "impress upon them the fact that the United States Government is able and willing to protect them and fulfill all its treaty stipulations."[40] To the people of the Cherokee Nation enduring food shortages, such a statement certainly meant annuities of food and clothing. Emphasizing the importance of these items, Pike warned his Confederate superiors that many Indian soldiers had not been paid in a year and were never issued uniforms.[41]

Another mounting problem was the increasing number of lawless, unattached men migrating to Indian Territory. Pike claimed that the region around Doaksville in the southern Choctaw Nation was overrun with white men who were avoiding the Texas draft, and Hindman claimed that families were fleeing the Cherokee Nation and northwestern Arkansas due to the prevalence of violent men he called "Jayhawkers, tories and hostile Indians." He added that the Cherokee region was "wholly exhausted of subsistence and forage," and that was in November, shortly after harvest.[42] While the promise of annuities was strong, protection in the form of law and order was also inviting. A spate of minor Union victories at Cowskin Prairie, Spavinaw Creek, and Locust Grove influenced the defection of Drew's Cherokee to the North. In all, around 1,500 new recruits flocked to Weer's command: some out of loyalty to the Union, some for food.[43] Cooper was convinced that the allied Indians were changing sides and joining the Union forces primarily because of their severe destitution and the inadequacy of Confederate protection.[44]

Pike blamed this defection on the Confederacy's inability to provide the basic requirements for surviving a prolonged drought, namely, clothes, money, and food.[45] To make matters worse, a Union detachment arrested John Ross at Park Hill before he could complete an order to conscript all adult males for Confederate military service to defend the Cherokee homeland. Precisely why Ross did not flee his home at the approach of Union troops is not known, but once he was in Kansas, the chief left for Washington to renew his tribe's allegiance to the Union. The whole affair smacks of collusion between the chief and the Union forces, but his arrest without resistance might also be seen as surrender. Then again, the fact that his captors simply released him to travel to Washington, D.C., appears all the more to be part of a deal negotiated for his defection. In his absence, the Confederate Cherokees elected Stand Watie as their chief and continued to conduct guerrilla raids against northern troops.

As the Union army moved southward, its problems intensified. Every mile marched into Indian Territory gave Confederates a better chance of severing vital supply lines to Kansas, while the parched earth added to the difficulty of procuring forage for the Union mounts. As a result, northern troops were often dispersed to enable easier local provisioning. Colonel Weer rushed his men to a spot 14 miles from Fort Gibson, but in doing so he outran his supplies. Here he camped for 10 days, 160 miles from Fort Scott, without resupply. The landscape offered no subsistence for man or beast; the available pools of water were stagnant and putrid. With only three days of provisions left, and no knowledge forthcoming of the whereabouts of the Union supply train, Colonel Charles E. Salomon, commanding the Ninth Wisconsin Volunteers, convened a war council of officers to convince Weer that to ensure his supply lines, he needed to move back toward Fort Scott. The next day, Weer placed the troops on half rations and refused to order a mass retreat. At this point, Colonel Salomon placed Weer under arrest and took charge of the Union forces.[46]

From this point on, the expedition fell apart, and Salomon initiated a withdrawal back to Fort Scott. The first Federal invasion of Indian Territory had come to an ignoble end. Still, Blunt took the situation in stride, informing his superior, Brigadier General John Schofield, commander of the Army of the Frontier, that as his army moved through Indian Territory, he planned to take all of the wheat, cattle, and hay he could find back to Kansas. Further, he would reward the loyalists by purchasing their produce and livestock and punish southern sympathizers by confiscating theirs, to make "certain that this country will afford a short living for a bushwacker when I leave it."[47]

Unfortunately, the activities of lawless men increased after the retreat of Union forces. Hannah Hicks, the daughter of Reverend Samuel Worcester and wife of Abijah Hicks, a progressive Cherokee who had bought the Park Hill mission and outbuildings in 1861, wrote a diary depicting the day-to-day dangers civilians experienced in a war-torn and drought-impoverished region, beginning shortly after her husband's murder on August 7, 1862, and ending February 20, 1863. The Hicks' pro-Union sentiment made them targets for rebel soldiers and secessionists in general. Confederate troops took her brother D.D. captive and cleaned out his store of all its medicines, leaving her to raise her three children without medical supplies. She reports murders perpetrated by Pins, Confederate sympathizers, and bushwackers, though she never ascertained for sure her husband's killer. Rebel soldiers confiscated her cattle and horses and even invaded her home, raiding "every closet, trunk, box & drawer they could find." They also took three barrels of sugar, blankets, and pillowcases, and tore the lace off of her bonnets. Even some of her personal letters wound up in a robber's possession, one supposes for his entertainment during the slow times at camp.[48] Those fortunate enough to have never experienced war may wonder how these men could steal food from a widow and her children, but such acts were common during the tensions of the Civil War, especially in Indian Territory.

The war had greatly reduced civilian access to manufactured textiles, and denim was exceptionally difficult to come by. In 1863 a yard of the durable cloth sold for the unbelievable price of seven dollars.[49] Hicks also relates two stories that reveal the high demand for clothing in Indian Territory: Once, a Rebel detachment captured a pro-Unionist, then stripped him of his clothing, giving him rags in exchange. No doubt the rags were what a Confederate soldier had been wearing before foisting them onto the man for his apparel. In another instance, bushwackers surprised two travelers and ordered them to strip on the side of the road. When one man fell to his knees and began fervently begging for his life, the criminals coldly shot him in the head. That was when the other traveler scrambled to his feet and took off running, making good on his escape and living to tell about it. The brigands obviously desired that the clothes of the travelers remain undamaged and unbloodied, thus ordering the men to disrobe prior to what would certainly have been both of their murders.[50]

The deprivation and competition for resources created by two years of war and drought resulted in a very thin line between the activities of soldiers and those of common criminals. Unfortunately, the war was not near completion, and neither was the drought. Conditions in Indian Territory would only worsen, forcing most to flee the region for safety and food. Refugees would make their way to Texas in the South or Kansas in the North, draining the resources from those areas.

NINE

The Last War Years in Indian Territory

*Drought is one of the best examples of our helplessness before
the broad-scale phenomena of nature.*

IVAN TANNEHILL

After the Union's withdrawal from Indian Territory in the summer of
1862, the war in Indian Territory had become a stalemate by the fall.
With Pike's command deprived of non-Indian troops, weapons, money,
and wagons, he witnessed the Indian Expedition from afar and knew there
was little he could do to stop it. Fortunately for him, the drought and
dissent from within doomed the first Union invasion of Indian Territory.
The Confederate forces in Indian Territory were also falling victim to the
malady of internal bickering. Cooper had ordered all of his men to move
south of the Canadian River to avoid being annihilated by the northern
forces, while Pike's men remained near the Red River. When Major General
Thomas Hindman, a strident secessionist from Arkansas, replaced General Earl Van Dorn as commander of the Trans-Mississippi Department of
the Confederacy, he found his native state in a woefully inadequate military condition. Most of its soldiers were fighting for the South from east
of the Mississippi River. Military provisions and armaments likewise had
been taken from the state for armies further east. Furthermore, the state
was penniless and could not purchase any of the needed materials to defend its borders. Hindman took drastic measures: He initiated a draft in
Arkansas, ordered the requisitioning of supplies from civilians, and even

authorized the use of guerrilla tactics against Union armies. He began looking for troops wherever he could find them, and hearing of the presence of
Confederate troops in Indian Territory, he ordered Pike to take a position
farther north. The eccentric lawyer-turned-general failed to act on these
orders and chose instead to offer his resignation.[1]

On July 31, 1862, Pike sent a circular to all of the tribes within his command. Explaining that the circumstances which his command now faced
were not of his design, but were the result of orders from district headquarters in Little Rock, he called on all the Natives to remain loyal to the
Confederacy and exonerated himself and President Jefferson Davis.[2] These
actions were highly suspect in military circles. His resignation had not been
approved, and the act of leaving his post was desertion. Further, his commanding officer had strong grounds to take issue with Pike's criticism of
Confederate war management and, coupled with this, Pike's issuing these
criticisms in a bulletin to his Indian troops smacked of insubordination.
When the circular made its way to Cooper's camp, the colonel suppressed
the copies and then wrote Hindman, stating that Pike was "partially de-
ranged and a dangerous person to be at liberty among the Indians."[3] Hindman agreed, and ordered Pike arrested and transported under military
escort to Little Rock, but fortunately for Pike, his resignation had been
accepted in Richmond.

Cooper now found himself in command of all of the Confederate forces
in Indian Territory, and he willingly complied with Hindman's orders to
bring his Choctaw and Chickasaw troops to the defense of Arkansas, even
moving into Missouri, where his units were involved in engaging Union
forces at Newtonia in September of 1862. Although the Rebel soldiers won
the day, they were forced to retreat the next month by the assembling of a
much larger Union army.

A new Union commander would try his hand at taking Arkansas. General James G. Blunt led a sizeable army into northwestern Arkansas and
defeated Hindman's Confederate forces near Old Fort Wayne on October
22, following that with another victory at Prairie Grove on December 7.
Capitalizing on these Union successes in Arkansas, Blunt ordered another
invasion of Indian Territory with the Indian Brigade, this time under the
leadership of Colonel William Phillips, a Scottish immigrant famous for his
antislavery newspaper articles covering the events of Bleeding Kansas. Phillips proved to be a good choice to replace Weer; he sent the Confederate
troops that had remained in Indian Territory fleeing south of the Canadian

River. As 1862 came to a close, Union forces once again occupied Cherokee Nation, taking Fort Gibson, burning Fort Davis across the Arkansas River, and threatening Fort Smith. Most of the Confederate troops remained below the Canadian River, and Pike's replacement, Brigadier General William Steele, took command of Indian Territory from his headquarters at Fort Smith.

With Confederate troops to the south of the Canadian River and Union troops occupying Cherokee Nation, pro-Union Cherokees took control of the tribe's political leadership. In the absence of Chief Ross, Thomas Pegg acted as principal chief and led the tribe in repudiating the treaty with the Confederacy, renewing its allegiance to the United States, declaring Stand Watie and his followers to be renegades, and ordering the confiscation of their property. With the Cherokee Nation seemingly pacified, the commander of the Army of the Frontier, Brigadier General John M. Schofield, a West Point graduate and science professor before the war, ordered Phillips to return the refugee Cherokee, Creek, and Seminole people to their homes in Indian Territory. Schofield also recommended that the Indians begin planting crops for the coming spring.[4] Such was the need for creating a source for provisions in the destitute region that General Blunt suggested using Union soldiers to help with the cultivation, if needed.[5] The combined effects of the drought and the abandonment of the Cherokee tribal domain had created a void of provisions for both armies. Schofield was determined that northern Indian Territory would provide fodder for cavalry mounts and wagon teams. As had become evident in the first two years of the war, supplies could not be hauled from Kansas to Indian Territory without this forage, making it nigh impossible to maintain a military presence in the region. The drought had also reduced the grass cover, reducing the health of the horses and mules for both armies. Phillips noted that the Union mounts had "all been used up," and that the stock was "very poor."[6] Due to the lack of forage, Confederate command ordered its forces in Indian Territory to relocate near the Red River to open up supply lines with Texas, leaving Fort Smith in a tenuous position.[7] Union troops occupied Fort Gibson, threatening Fort Smith from the west, and more Union soldiers were poised to move on Little Rock to the east.

Simply stated, the drought and the resulting dearth of subsistence limited the actions of both armies and brought severe hardship on those who attempted to survive in the region. With the Arkansas River too low to allow steamboat traffic to Fort Smith, the Confederate soldiers stationed there suffered a dire lack of provisions. The depleted stores of hay and corn at Indian Territory military posts meant that teams pulling wagons from Texas could not travel to or from the stations. Due to the lack of forage, most of the provisions being transported in supply wagons from Texas to Fort Smith were consumed by oxen on the way, leaving no feed for the return journey to Texas.[8]

Even in Union-held territory, the destitution was great. Colonel Phillips ordered the distribution of flour and cornmeal to starving families in the Cherokee Nation; some of the women claimed they had not eaten bread in two days. The Union army suffered the same problem with feeding the teams that pulled their supplies. Echoing Pike's assessment of the Union's chances of success in invading Indian Territory the previous year, Phillips took solace in the notion that the same factors would impede his enemy, claiming "the great scarcity of forage renders it difficult for them [the Confederate forces] to enter or travel over the country."[9] In early February, Phillips sent another wagon train of provisions to around a thousand starving denizens of Fort Gibson, where the mills remained dormant due to the absence of grain.[10] This strategy of feeding the destitute in Cherokee Nation daily won over more and more tribal members to the cause of the North.[11] Few rewards motivate loyalty among a starving people like full bellies. Phillips suggested including the Creeks and Choctaws in the rationing as well, because, if not, these Indians would be fighting against him in the coming summer. As Phillips aptly put it, clothing and feeding Indian men and their starving families was "cheap recruiting."[12]

Animals experienced the effects of the drought as acutely as humans. While camped near Pea Ridge in Arkansas, Wiley Britton observed that horses tied to trees had gnawed the bark off the trunks as high as they could reach and that they were so hungry that they chewed the manes and tails off other horses.[13] Watie, struggling to to destabilize the Union-held Cherokee Nation, was reduced to yoking up cattle to his supply wagons in the hope that they could perform like oxen, because the supply train mules were dying of starvation. To save his unit's horses, Watie ordered them driven south into Texas and, in the process, left his men without mounts, limiting their ability to conduct guerrilla raids into Union territory.[14]

The months just prior to the harvest of wheat in June and corn in September proved exceedingly difficult for those residing in Indian Territory. The supply from the previous year's low yields vanished, leaving people to scavenge off the countryside or steal from neighbors. Unsurprisingly, crime increased dramatically. Steele warned Cooper that large groups of deserters and outlaws were menacing the Arkansas River valley. One such band of "bushwackers" murdered and robbed up and down the Arkansas River under the leadership of Martin D. Hart, a Texan who claimed to be commissioned as a captain in the First Regiment of Texas volunteers.[15] A more notorious episode of the nonmilitary violence occurred when a man named Benge led a mixed party of whites and Indians on a murdering spree. They dragged a fifty-year-old man from his home and killed him, along with a sixteen-year-old boy who was suffering from smallpox.[16] Steele complained, "there is scarcely a day that I am not in receipt of some sad tale of murder and outrage." He blamed the Confederacy for ineffectively maintaining law and order and for not providing food to the starving victims of the drought, both of which had a direct impact on the crime rate.[17] He obviously had no idea of the extent or duration of the dry trend.

The violence led some citizens to disregard the conventions of warfare: a group of Rebels wearing Union uniforms killed seven Pin Indians near Park Hill in late March.[18] Desertions from the Confederate army increased as living conditions worsened and men left their posts to attend to their families. The lack of adequate nutrition increased the roll calls of sick bay. The desertions and illness lowered morale and reduced the Confederate fighting force in Indian Territory from 40,000 to 18,000 in the spring of 1863, making any planned counteroffensive moot.[19] In an effort to ensure a food supply, Steele sent his chief quartermaster, Captain Cabell, to Bonham, Texas, with orders to personally oversee the purchase and transport of supplies because, in the words of the commander, "if the mountain would not come to Mohammed, Mohammed must go to the mountain."[20]

The worsening environmental circumstances and rising violence were taking their toll on the Native people of Indian Territory. Some of the Muscogee and Choctaw people who had thrown in their lot with the Confederacy began to question the wisdom of their actions. Moty Kinnaird, Echo Harjo, and other Creek principal chiefs wrote to President Jefferson Davis, listing their grievances with the Confederate States of America and pointing out that the Creek tribe had surpassed its treaty obligation to raise one regiment for the protection of Indian Territory by raising two. Further, the Confederacy had promised to provide uniforms, weapons, and ammunition, but often these necessities were siphoned off to white units, which left the Indian regiments underfed, underclothed, and underarmed.[21]

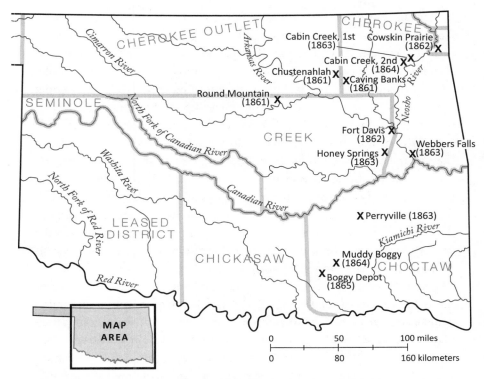

Civil War Battle Sites in Indian Territory. Map by Tom Jonas. Adapted from a map in *The Story of Oklahoma*, revised 2nd ed., by W. David Baird and Danney Goble. Copyright © 2016 University of Oklahoma Press.

The Choctaw government had sent Elias Boudinot to the Confederate capital in Richmond, Virginia, in an attempt to procure famine relief and military materiel, but despite his best efforts, he was unable to obtain any financial resources during the winter and spring months of 1862–1863.[22] In fact, the Confederacy expended few resources in Indian Territory, for they had more pressing problems east of the Mississippi River. General Lee was moving his army northward into Pennsylvania, and Union forces under General Grant were closing in on Vicksburg, the last Confederate-held point along the Mississippi River. Once this city fell to Union forces, the task of bypassing the Union blockade on coastal trade would become much more problematic, as the current solution was to move medicines and weapons imported from Mexico into Texas, then freight them through Arkansas or Louisiana and across the Mississippi River.

Reminiscent of the Comanche survival strategy earlier in the drought, Confederate Indians, who were to a large degree fending for themselves, turned to raiding Union livestock. Cooper moved his army of five thousand men northeast across the Canadian River and took a position just five miles from Fort Gibson. At nine o'clock on the morning of May 14, 1863, Cooper and his men attacked the federal unit guarding the cattle herd, but failed to drive off the animals.[23] Less than a week later, the Rebels made off with a majority of the Union livestock, a double benefit to Confederate forces that fed their own soldiers while withholding protein from their enemies.[24] The loss could not have come at a worse time for Phillips, whose daily rations had already been reduced to two ounces of flour a day. This additional deprivation greatly diminished his troops' effectiveness and, according to the colonel, limited his ability to recruit Indian troops.[25] Blunt organized a supply train of three hundred wagons and sent nine companies, some artillery, and the newly formed regiment of the First Kansas Colored Infantry to guard it. Stand Watie knew there was only one trail that could accommodate the travel of such a large group, and he followed the supply train's daily movements down the road through informants. The wily Confederate planned to attack the wagon train at the ford on Cabin Creek with two thousand men. On July 2, 1863, just as Lee's army prepared to march on Union General George Meade's flanks at what became known as the Battle of Gettysburg, Watie engaged the Union supply train in far northeastern Indian Territory. Here, Union troops drove Watie's men from the ford and continued their mission to Fort Gibson. The wagon train limped into Fort Gibson on July 5, 1863, ensuring the continued Union occupation of that fort and of Cherokee Nation.

General Blunt then led another group of Union reinforcements to Fort Gibson and made plans to invade Indian Territory south of the Canadian River. Facing the growing force of Union troops at the fort, Cooper moved his six thousand men to Honey Springs, twenty-five miles south, and Steele planned for General William Lewis Cabell, commanding three thousand Confederate troops in Arkansas, to join Cooper there. Honey Springs was a critically strategic location. First, there was dependable water there in the form of several springs, and second, the Confederates had made it their supply depot. Blunt decided to strike Cooper before Cabell could join him and perhaps take the Confederate supplies in the process.

Summer rains returned in 1863, and Blunt's men slogged through muddy roads, probably thankful for some relief from the heat. On July 17, the two forces met in what would become the most decisive Indian Territory battle of the Civil War. Confederate soldiers were hampered by the poor quality of their gunpowder, which became useless when dampened to the least degree, and as fate would have it, the scene of the battle was overcast and drizzly all day long.[26] The Confederate troops were forced to retreat, but, rather than let their supplies fall into Union hands, the demoralized Confederates burned all they could not carry.[27] Blunt did not remain south of the Arkansas. Instead, he marched his troops back to Fort Gibson to enjoy their victory, obtain fresh supplies, and await additional reinforcements.

The Rebel soldiers' frustration reached the breaking point that summer. A rumor flying around Rebel campfires accused the Yankee high command of sending an agent to Mexico to sell faulty gunpowder to Confederate purchasers.[28] This scuttlebutt seemed likely, for the powder looked good when the cans were opened, but the slightest dew turned it into paste.[29]

In July, rains swelled the Arkansas River to ten feet above its previously recorded highest stage, hindering further Union advances but giving hope for the coming harvest.[30] The dry conditions likely exacerbated the flooding. The parched soil hardens under the baking of the sun, and when the rain does fall it quickly runs off to the nearest creek or river. Not surprisingly, Confederate desertions increased after the Battle of Honey Springs as men went home to help gather in the crops. Steele was alarmed to hear that a company sent out to overtake deserters found a party of two hundred headed by an officer and bound for Arkansas.[31] Such large scale-desertions were not uncommon and further represented a contempt for military protocol.

Blunt was not in Fort Gibson long before the awaited reinforcements arrived, and the general was keen to resume the campaign against the Rebels. Bolstered by the news that Little Rock had fallen to Union forces on September 10, 1863, he moved south with close to 4,500 soldiers, knowing that Steele had taken command of the troops in the field. Steele coopted Cooper's units as well and, hoping to divert Blunt's attentions, sent Cabell back to Fort Smith, all the while awaiting the arrival of the Texas home guard led by General Smith P. Bankhead. As Cabell traveled to Fort Smith to take charge of that post, Bankhead, fearful to stray too far from Texas, refused to go further north than Boggy Depot in southern Indian Territory, claiming that the top priority should be protecting the remaining supply stores.[32] Steele was in a difficult position, and his troops' morale continued to plummet. Cabell's soldiers and Cooper's Indian troops took long walks nightly to their Arkansas cabins, realizing their weapons were useless with the pasty gunpowder, and now Bankhead was reluctant to move north, which was where Steele wanted to make a stand. Steele understood that the Creek people would soon be lost to the North, like the majority of the Cherokees, if he could not demonstrate vigorous defense of their homeland. Finally, at Perryville, Steele's men gave a halfhearted defense of the road to Boggy Depot then fell back, leaving yet more supplies for Blunt to burn. Here, the Union general turned back toward Fort Smith, hoping quick action on his part might wrest the post from the Confederacy. The result was anticlimatic: Cabell withdrew from Fort Smith, leaving 600 sick and wounded who could not travel and losing another 13,000 bushels of corn to Blunt's force.[33] With the fall of Little Rock in September of 1863 and the difficulty in transferring supplies overland from Texas, holding the post was simply impossible.

The war had moved south into Choctaw country, and now these people had to find ways to cope. The civilians of Gaines county in the Moshulatubbee District of Choctaw Nation began pushing their livestock up into the mountains to safeguard them from the approaching Union army.[34] The people of northern Choctaw Nation were desperate for food and clothing. Local theft of victuals was common, as Alfred Wade, the first governor of the Choctaw Nation, commented in a letter to Peter Pitchlynn: "there is more injury done to the people by Southern people th[a]n by [the] federal [armies]."[35] Wade also noted, while crossing Sugar Loaf Mountain in modern-day Bryan County, Oklahoma, that the suffering of the civilians was acute. He saw that "some children and women [were] naked[,] nothing to buy corn with because their corn was taken by the federalists."[36] A few straggling Natives approached Watie's camp, hopeful for something to eat. The Confederate Cherokee was all too happy to give them food, but they had to work for it by doing odd jobs for Watie's unit.[37] As 1863 came to a close, much had changed. Union forces now held both Fort Gibson and Fort Smith, and the Confederate army was pinioned far to the south in Indian Territory, demoralized and frustrated but not done in just yet. As if to punctuate this reality, Stand Watie led a raid north of the Canadian River all the way to Park Hill, Cherokee Nation, where he ordered the burning of Chief John Ross's home. Still, such an act of bravado could not deter the hundreds of Confederate refugees from Cherokee and Creek lands who moved toward the Red River in the wake of Union occupation of their homeland.

At his wits' end as to how to defend Indian Territory with troops who tended to leave for home at will, a shortage of provisions and ordnance, and faulty gunpowder, Steele tendered his resignation in December. He went to Austin, Texas, where he submitted a report on February 15, 1864, of the previous year's activities. Steele notes the impact of the drought on morale, claiming "the scarcity of forage and subsistence, together with the destitute condition of the command encouraged desertion."[38] The parched landscape affected the military strategies of both the Union and Confederacy. Steele informed his superiors that Union forces were scattered over large sections of land because the drought-afflicted environment could not support large congregations of troops, but that he could not take advantage of this otherwise tempting situation because he was compelled to disperse his cavalry for the same reason.[39] Furthermore, his plans to attack Blunt's army were foiled when Cooper related to him that the horses had been without forage for several days and were in no condition to march.[40] Finally, the necessity of feeding thousands of indigent refugees who, if not cared for, would turn to the Union, reduced the amount of supplies for the Confederate soldiers, exacerbating his desertion problem.[41] Steele's replacement, General Samuel B. Maxey, was transferred from the Army of Tennessee to oversee a more aggressive campaign in Indian Territory.

The Union command also experienced a change, for no sooner had Blunt moved his headquarters from Fort Scott to Fort Smith than he received word of his displacement. Blunt's political enemies had outmaneuvered him and arranged for his replacement by John McNeil, who moved to the headquarters on the Arkansas River and inherited all the supply problems that the Confederates had encountered there. During the winter of 1864, the drought-shallowed river froze over, stalling water traffic, and the lack of transportation made hauling supplies from Fort Scott impossible. The absence of local provisions meant McNeil had to place his men on half rations or risk running out of bread within the week.[42] If these conditions continued, he too might have to abandon Fort Smith, as the Confederates had.

The situation was not as bad at Fort Gibson, which had been in Union hands longer and thus had received supplies prior to the onset of winter. In fact, Colonel Phillips planned a foray into Choctaw lands in which he would practice a "carrot and stick" approach. He disseminated Lincoln's pronouncement of amnesty to all Indians who would gather under the Union banner, but his expedition was punitive as well. In Phillips's words, "I shall leave no subsistence for a rebel army, or forage, so that all its supplies must come from Red River in any movement to the Arkansas, the stretch being 180 miles." His aggressive romp through Choctaw lands disheartened the Confederacy's Indian allies, for it proved that Maxey and his Confederates could not make good their promise to protect them.[43]

The danger to the Confederacy was real as its Indian allies were forced to arm themselves by taking guns from the Union dead. Maxey confessed to Kirby Smith, his immediate superior, that the same reasons that had compelled the tribes to ally with the Confederacy might now induce them to side with the Union: the absence of annuity payments and the withdrawal of troops, which, to the Indians, meant the removal of protection.[44] Fortunately for Maxey, Union troops destroyed the lure of amnesty by outraging Confederate Indians, killing eleven Seminoles in a firefight and mutilating their bodies, slitting their throats from ear to ear. Then, on Phillips's march, Union troops fired into a camp of fugitives, killing a ten-year-old child and a woman. Demonstrating the typical racism of the time, Maxey was shocked that these inhuman outrages were committed by white troops.[45] In May, Pin Indians killed seven civilians in Cherokee lands and stole flour and cornmeal at will.[46] These acts stiffened the resolve of many Confederate Indians, even though they endured depredations from Confederate deserters as well, for these were men desperate to feed their families. Watie wrote to his wife that "no property is safe anywhere. Stealing and open robbery is an everyday occurrence."[47]

The events of spring and summer 1864 helped to revitalize the Confederates in Indian Territory. Union forces set out from Little Rock and Louisiana in a two-pronged invasion of Texas. Sterling Price, the new Confederate commander of Arkansas, called on Maxey to repel the Union's southward march. Unlike Pike, Maxey held no aversion to asking the conscripted Indians to leave the confines of Indian Territory in aid of the rebellion. However, the Choctaw Second Cavalry Brigade under Confederate commander Tandy Walker was the only Indian unit to cross the Arkansas border and participate in the Battle of Poison Springs on April 18, 1864, in which the Rebels captured two hundred Union wagons loaded with forage, and helped turn back the Union advance. Then, in mid-June, Watie's men seized a Union steamboat carrying supplies to Fort Gibson. This small victory heartened the men, for they seized 150 barrels of flour and 13,000 pounds of bacon, but most of the men with homes nearby left immediately after capturing the vessel, carrying what flour and bacon they could to their destitute families and leaving only a skeleton force to deal with northern

troops, should they come. Sure as rain, a small detachment of Union soldiers appeared in time to threaten the stores that Watie's men had hauled to
the nearby bluff, forcing them to burn the boat and the remaining supplies.
The rising water swept away any remains of the charred vessel down the
river, leaving nothing to mark the Confederates' short-lived success.[48] These
events were only two small victories, but they bolstered the morale of those
supporting the rebellion in Indian Territory.

Maxey, seizing the momentum of these achievements, attempted to
strengthen the determination of those Indians whose loyalties straddled
the fence. In a circular dated June 15, he denounced the "cowardly skulks
who are deserting their comrades and country" and targeted the "low down
skuffs" who abetted them. Demonstrating an incredible gift for verbal imagery, he wrote that he wanted all the military personnel in Indian Territory to think of desertion as a "'stink ball' in the nostril of every soldier."[49]
Maxey continued his campaign to reduce the number of soldiers going
absent without leave by circulating announcements of the courageous acts
of Confederate Indian units, who had decided to make a statement to their
comrades by re-enlisting for the duration of the war. On June 23, 1864, the
men of the First Choctaw Brigade announced their intentions to re-up, and
four days later Stand Watie's Cherokee regiment followed suit.[50]

Feeling confident about countering Lincoln's amnesty offer, Maxey fo-
cused his attentions on the vulnerability of Fort Smith. The drought continued to make hauling supplies overland nearly impossible. The Confederate
commander understood that control of Fort Gibson was reliant upon con-
trolling Fort Smith, and both required the ability to have supplies transported in. Control of the Arkansas River was the most expedient means to
accomplish this.[51] In the summer of 1864, the Confederate government in
Richmond had sent an inspector to evaluate the supply situation in Indian
Territory. The individual designated for this project was Colonel and Assistant Inspector General E. E. Portlock. After inspecting the department,
Portlock wrote to Maxey: "What are our resources? No one, except from
personal observation, can form an idea of the utter destitution of the Indian
Territory except a limited supply of cattle."[52]

Seizing on this reality, Maxey encouraged the proven strategy of guerrilla
raids to cut off supplies to Union forces and remove the one commodity
that remained near the forts, cattle. In late July, Brigadier General Richard M. Gano routed a federal unit five miles from Fort Smith, taking 127
prisoners, but, more importantly, obtaining two hundred Sharps rifles, four

hundred six-shooter pistols, horses, sutlers' stores, and camp equipage.[53] On August 2, Cooper attacked a picket station outside Fort Smith, capturing twelve men, some horses, and cattle just three miles from the Union post.[54] These Confederate actions forced the area's civilian population to seek refuge within the walls of Fort Smith, straining the food supplies of the post and mandating the reduction of rations to half the usual supply.[55]

With 1864 being an election year, the Confederacy hoped for Lincoln's defeat by Democrat and retired Union general George McClellan, who ran on a platform calling for the cessation of hostilities and a negotiated settlement of the conflict. With the growing unpopularity of the war in mind, Confederate officials supported Major General Sterling Price's invasion of Missouri, targeting Saint Louis with its "supplies and military stores," and his retreat through Kansas and Indian Territory, "sweeping the country of its mules, horses, cattle and military supplies of all kind."[56] Hoping to divert the Kansas troops' attentions, Watie was to conduct raids in northern Indian Territory and, if possible, southern Kansas. While on this mission, the Cherokee officer conducted another attack on the vital supply train from Fort Scott to Fort Gibson, once again at Cabin Creek ford. This time his plan succeeded, and the whole train of 250 wagons fell into the hands of Watie and Gano, who loaded all they could on the most serviceable wagons and transported them to Tulsey Town. Here they issued clothing captured at Cabin Creek to two thousand of Gano's men and eight hundred of Watie's. Maxey claimed this victory was "a perfect godsend as the command was literally ragged."[57]

Although Price was unable to take Saint Louis, he did claim a few Confederate victories before suffering a series of defeats that drove him into Indian Territory, where he gave Fort Smith a wide berth as he passed to its west and made his way to Texas. An officer in Price's army wrote that they "endured the severest privations and sufferings during the march through Indian Territory to Boggy Depot." He went on to detail the trip: "for twenty-three days our animals were without forage. For twenty-three days we subsisted on beef without salt . . . and for three days were without food at all."[58] To make matters worse for the Confederates, Lincoln was re-elected, regardless of the unpopularity of the war. Peter Pitchlynn, the elected principal chief of the Choctaws, detailed the situation in his region as 1864 came to a close: "The information reaching me from every county and almost every neighborhood in this nation, representing a state of destitution unprecedented in our history. . . . In consequence of a large number of producers being engaged as soldiers and also the drought which prevailed last summer, very little corn was made in the uplands. The scarcity has caused all classes to look to the few farms of slaveholders on Red River as their only hope of subsistence until another crop can be produced."[59] The Choctaw principal chief went on to state that refugees, soldiers, and their families were too numerous for the available food supply. As a result, "soldiers, regular and irregular, are constantly traversing the highways and byways of the nation, taking by force or threats of violence the little that is to be found."[60]

The last year of the war was marked by suffering and corruption as competition for resources became acute over the winter and spring months of 1865. Samuel Maxey urged residents of north Texas to plant every spare acre in corn to alleviate the famine caused by the steady arrival of Indian Territory refugees and the absence of many farm laborers who were serving in the military.[61] Forced by the drought to address the food shortage, he later directed Watie to reduce his men to half rations of breadstuffs, as flour was in short supply, with months to go until the wheat harvest.[62] Making matters worse, Cooper had to inform Chief Pitchlynn that he could give the desperate Choctaws only 1,500 instead of the promised 2,500 bushels of corn.[63]

The drought caused conditions just as critical in northern Indian Territory. Fort Smith had attracted close to five thousand of its own refugees, who sought any relief offered by the Union installation. Loyal Creeks, Seminoles, Cherokees, Choctaws, and Chickasaws congregated within the shadow of the post for protection and rations, but in late March they were on the verge of starvation, having done without any supplies for three weeks. The same problem that had plagued the fort all during the war continued to make remaining there difficult. The question became how to get supplies in when the river was low. Teams required feed, and there was none in the environs to supply draft animals for a return journey to Fort Scott or elsewhere.[64] Union commanders of both Fort Smith and Fort Gibson were at a loss as to how to feed the starving Natives as well as their troops.

As news reached Indian Territory of General Lee's surrender at Appomattox on April 9, 1865, the main concern in Indian lands was the rising tide of crime. Cooper warned his superiors that "There is more danger of anarchy from lawless bands of armed men in case the war is stopped than there is now from the public enemy [Union troops]." Cooper echoed the fears of Indian Territory citizens, which were turned from the soldiers of the opposing army to the "lawless and desperate men of our own [army] who are beginning to feel free from all restraints."[65] Starving people will typically obtain food any way they can. The situation at the close of the Civil War was deadly. Armed men with starving families and no legitimate way to raid for food since the hostilities had ceased resorted to whatever means they could to obtain sustenance. Watie thought it best to send most of his soldiers home on furloughs because "hints have been thrown out that they would help themselves to public property."[66]

On May 26, 1865, General Kirby Smith surrendered as commander of the Trans-Mississippi Department, but he lacked the authority to surrender the Indian fighting units, as the tribes, acting as sovereign nations, had each made their own treaties with the Confederacy. The Chickasaws and Caddos surrendered on June 14; Chief Pitchlynn capitulated for the Choctaws five days later; and the last Confederate army commander to surrender was Stand Watie, who finally did so for the Cherokees, Creeks, and Seminoles on June 23, 1865. As these men returned to their tribal homelands, they too became part of the starving populace. James C. Veatch, the commander at Fort Smith, noted that the number of people moving toward Fort Smith immediately after the Civil War grew exponentially. He wrote to his superiors that the government would need to feed about 9,000 people near his post alone, or the destitute war refugees would take what they needed by force. Citizens of north Texas had sent charitable donations of food, but there simply were not enough resources in the region to satisfy the massive hunger.[67]

At Fort Washita when Cooper's Texas troops found out that General Smith had surrendered, they "disbanded and appropriated most of the public property," including supplies of food, horses, cattle, and wagons. They took anything that would help their families survive post-war starvation.[68] General Blunt, who had returned to command Union troops in Kansas, was also worried about the Indian troops after the war, for they returned home with their guns. He wrote to Colonel John Levering, adjutant general of the Department of the Arkansas, that "since mustering out of the Indian regiment there is a disposition manifested among them to organize in small bands, for the purpose of thieving and preying upon the country, and they do not hesitate to kill whenever necessary to accomplish their ends."[69]

After four years of outmigration from Indian Territory, the end of the war brought a return of the Indian refugees. Sadly, the one food resource that Indian Territory was still producing, cattle, had become the target of criminal activity and military corruption. The strategic war for beef rampaged across the tribal domains as each army tried to commandeer the supply. As early as the spring of 1863, General Blunt was convinced that Rebel soldiers were stealing U.S. Army cattle and herding them from Indian Territory in the direction of Pike's Peak.[70] One wonders just what he thought the Confederates would do with herds of cattle in southern Colorado. A more likely objective of the drives might have been the Indian Reservation at Bosque Redondo along the Pecos River in New Mexico. By 1864 trade in contraband beef was burgeoning among Union military personnel. Civilian "bushwackers" stole cattle in Texas or Indian Territory and sold them to contractors, who then sold the beef to the U.S. Army.[71] Milo Gookins, the Wichita Indian agent, claimed members of the Union cavalry were also rustling cattle for sale to the contractors.[72] Federal authorities saw no way to control the black market trade other than to confiscate all cattle brought up from Indian Territory and forbid any permits to contractors for purchasing beef south of Kansas.[73]

At the war's end this problem only worsened. U.S. Creek Indian agent George Cutler reported rustlers had been driving large numbers of cattle out of the tribal domain into Kansas, selling them to contractors under the pretense of having written authority from the Creek tribe. Cutler informed Major General Blunt that these cattle were, in fact, stolen, and that no such documents existed.[74] Agent Gookins claimed the cattle ring was bribing U.S. officials to turn a blind eye to these goings-on, and even accused one General Ford of accepting these bribes.[75] The superintendent of Indian affairs put the number of stolen cattle during the war years at 300,000 in his annual report to Congress.[76] The contraband trade also illustrates the catch-22 that many reservation Indians faced at this point in our nation's history. Rustlers were stealing their cattle, and no redress was forthcoming from the federal government. When the Natives tried their own hand at cattle thieving to feed their families, they were prosecuted. C. C. Snow, in charge of the Neosho Agency in Kansas, reported that some Seneca and Quapaw men resorted to rustling because they had not been paid for their enlistments, and they either "must starve or steal."[77]

The war era in Indian Territory proved General Sherman's dictum that "war is hell." The suffering created by drought conditions before the first shots of the Civil War at Fort Sumter was minor compared to the ensuing hardship brought on by untended farms and relentless raiding by both armies during the war. As the refugees returned to Indian Territory, all was not singing birds and sunshine; families had to cope with continued food shortage, increased criminal activity, and government corruption. Also, the tribal decisions to side with the Confederacy came with a high price. Eighteen sixty-five was a watershed year for the Five Tribes, for it demarcated the end of an era when tribal governments acted on their own policies; for the next hundred years or more, they would primarily react to white policy.[78] In the last month of that epic year, an act of altruistic kindness shed a ray of hope for the future. After all the killing, stealing, and profiteering, Cyrus Kingsbury, a beloved missionary to the Choctaws, sold off much of his property to assist the needy, and by the end of the Civil War there were all too many needy in the Choctaw Nation.[79]

TEN

Crime in Civil War Texas

*No human power or agency is adequate to battle successfully
against nature and the elements.*

MAX KRUEGER

While the Civil War years devastated Indian Territory, the buffer between Kansas and Texas, the war years in Texas exaggerated pre-
existing problems there. As if the Indian raids were not enough, domestic
crime continued to threaten to the stability of the region. The pre-war
years, marked by outmigration, were followed by significant immigration
in the war years. Due to its distance from the military fronts, the lack of
an effective police force, and its proximity to ranches and their cattle, the
frontier region of Texas attracted all sorts of ne'er-do-wells, including draft
dodgers, deserters, northern sympathizers, and criminals. And since that
influx of people occurred during the prolonged drought, competition for
resources was even fiercer.

As we have seen, the antebellum years on the Texas frontier were marked
by violence due to that struggle for resources. A September 1858 story in
the *Dallas Herald* reported that, within a two-week period, thieves robbed
forty thousand dollars from the mail deliveries between San Antonio and
Galveston.[1] Another violent encounter described in a newspaper article that
same month was instigated by a struggle for control of a water source and
led to the deaths of four individuals. A man by the name of Crook had
one pond that still held water on his property near Sulphur Springs; all the

others in the area had dried up. As Crook's pond was the only source of the precious liquid for miles around, neighboring ranchers pushed their cattle to it, drastically reducing its stores. Crook contacted one neighbor, a man named Musgrove, and let him know that neither he nor his cattle were welcome to visit Crook's watering hole. Not intimidated, or perhaps made fearless by necessity, the unwelcome guest, accompanied by three others, came across the open pastures the next day, driving the neighbor's cattle toward the lone pond, but this time Crook and three of his own comrades met them midway. All six men began firing as soon as they were within range, and two from each party were mortally wounded, including Crook's son.[2] The fight to control water had become deadly.

On the eve of the Civil War, communications on the frontier of Texas had been extremely limited. Railroads had not yet crossed the region, and wagon roads were primitive. Vegetables spoiled before they arrived at major markets, and the price to freight crops like wheat and oats was too high to justify the cost of raising them. With the region in the throes of a major drought, growing crops was an extremely unpredictable venture, and cattle became the only viable trade commodity for people remaining in the area. As early as April of 1857, John Baylor had written that stock raising was the only endeavor that "would pay bills in Texas."[3] Livestock was also a form of moveable wealth. If a market was not nearby, a stockman and his horse could push the cattle to it. It is not surprising, then, that more thefts of that moveable moneymaker were reported during an era of extreme drought and fewer regional markets.

In many instances, frontier settlers blamed the cattle thefts on hostile Indians who had indeed raided into Texas for more than one hundred years, but as we have seen there are a few serious flaws in such a blanket accusation. During the first half of the nineteenth century, Plains Indian raids targeted horses and captives, which could be traded for desirable items. Captured men might also become tribal horse herders and perhaps warriors and hunters, while women could be taught to tan buffalo hides if not ransomed. To a nomadic society whose economic livelihood lay in the hide trade, tanning leather was an important function. Killing a buffalo, while dangerous, took little time; processing the hides was another matter. Pulling the tough skin from the muscles was arduous. The hide then had to be staked out on the ground with the inside facing the sun, so that Native or captive women could remove any fat by rubbing the hide with buffalo brains and a scraper, making the hide soft and supple. Finally, the hide was stretched over a fire to give it the added value of water resistance. The average skilled female tanner could process ten hides in a year, but a skilled Native could prepare more than eighteen.[4] Thus, the Indian raids were far more than acts of terror intended to curb non-Indian settlement. They were also vital to the Native economy. When the ability to hunt, particularly after the introduction of the horse and firearms, outpaced the capacity to process the hides, many hunters and warriors participated in raids to obtain more women to supplement their labor force, eventually enabling them to have more hides with which to barter for trade goods. Thus horses and women were preferred over cattle as objects for raiding.[5]

Another reason Indian raids generally ignored the growing number of domesticated cattle during the early 1800s was that cattle moved too slowly for raiders to escape quickly with them, and beef was not the preferred meat of the Comanche and Kiowa anyway. Thus, many raiding parties, disgusted by the burgeoning number of domestic bovines that diminished the forage of the range, simply shot any cattle they passed. Rachel Parker Plummer, a victim of the infamous Comanche raid on Fort Parker, recalled that the Indians "killed a great many of our cattle as they went along," toting her off to captivity.[6] Like the fisherman who never swerves his car to avoid the turtle, a fishing competitor, the Natives understood that cattle competed with bison for grass and represented an undesirable European transformation of the ecosystem's flora and fauna.

At any rate, by the close of the Civil War, the Comanches' opinion of beef had reportedly changed. Citizens of Lampasas County claimed that Indian depredations "extended to cattle, and [were] not confined mostly, as heretofore to horses—This shows the audacity of the parties and the impunity with which they expected to carry on their robberies."[7] It also exhibits a new necessity. As the buffalo migrated out of the southern plains in search of water during the mega-drought, the Plains Indians needed a new source of meat. Cattle, although not as desirable as bison, could fill that dietary niche, and the presence of non-Indians in Kansas provided a market for those animals not utilized as food.

As the Civil War preoccupied most Native and non-Native Americans' thoughts, the violence on the Texas frontier mounted. During the early years of the war, petitions continued to pour in to the governor's office, pleading that local militia units be raised on the frontier. Gatesville rancher John Chrisman, after witnessing an Indian raid, called for the formation of a local minuteman unit at state expense, in the hope that it could protect his and his neighbors' property.[8] Alexander Walters, also of Gatesville, echoed these requests: "We are desirous that publick arms should not be taken from us, we are very thinly settled in this portion of [the] country and are mostly poor people and have been Engaged for the last twelve or fifteen months in running after Indians trying to get our stolen property from them so that most of us is unable to buy arms and if we were disarmed the only alternative would be to leave the country and go further East which would result in the ruining of many."[9] The phrase of interest here is "we are a poor people." Surely they were poor and desperate too, with a severe drought burning up the crops, the removal of military posts and Indian reserves and their accompanying beef markets, and the depletion of their herds by rustlers and Indians. The creation of local militias would provide a small income for area men while giving them a better chance of protecting their investments.

Oliver Loving, a rancher and store owner from Palo Pinto County, came up with a novel solution to the absence of markets after the Texas Indian reserves were closed. He had heard of the gold rush in Colorado. All those miners needed something to eat, and the successful ones would be certain to have the financial wherewithal to purchase a steak. He drove what remained of his herd to Denver in 1860, but before he could begin his return trip to Texas, the Civil War broke out. Union authorities refused to allow him to leave until Kit Carson interceded on his behalf. In 1862 a miffed Loving wrote to Texas governor F. R. Lubbock, claiming that Union officers were paying Comanche warriors for Texas scalps.[10] This report was certainly an exaggeration, but there is evidence that Union officers encouraged Natives to raid into Texas, hoping to preoccupy the Confederates and limit their ability to participate in the war.[11] There was the added bonus that every cow taken from the Texas frontier would end up somewhere other than a Confederate belly. The moveable wealth of the herds became a military target in Texas as well.

Whether at the encouragement of Union officers or as a means to turn back non-Indian settlement and obtain much-needed trade items and stock, Comanche and Kiowa war parties continued to focus their attentions on the Texas frontier during the Civil War. J. Y. Dashiel, the adjutant and inspector general for the State of Texas, claimed that "the country [is] absolutely full of Indians, some of them have Guns, but for the Most part bows and arrows. They are doing Great mischief! Consternation pervades the entire frontier which is constantly receding."[12] Such reports were common during the war years, yet they can be misleading.

Blaming all cattle thefts during the war years on Native people alone is a convenient assumption, but one that can be shot full of holes upon close scrutiny. Brevet Major General Persifor Smith, a Mexican War veteran who took command of the Department of Texas in 1851, was convinced that the newspapers' practice of borrowing stories of attacks from other publications but treating them as new events exaggerated the number of Indian livestock thefts. Thus, an original attack that caused the deaths of four frontiersmen could quickly be spread throughout Texas's news circulars, the number of casualties growing with every repetition.[13] Another source of criticism lies within the accusations themselves.

Outlaws and, later, military deserters from the war placed extra pressure on area settlers and cattle ranchers. Non-Indians and Indians both could profit from the cattle trade, and if the rustlers could find a way to blame their thefts on Native raiders, so much the better. Of course, blaming Native people for non-Indian crimes had been going on for a long time. European settlers had donned Indian costumes to conceal their identities at least since 1734, when Lord Dunbar and his men were attacked by black-faced and feathered hostiles as they attempted to enforce the Mast Tree Law in New Hampshire.[14] The attackers proved to be non-Indian settlers who resented the law and let their displeasure be known violently.

The practice of falsely setting up Indians as the perpetrators of white crimes, now more than a century old, had become commonplace on the Texas frontier. In 1857 a group of U.S. soldiers disguised themselves with Indian dress and murdered one of their officers at Camp Cooper, intending to blame the reservation Indians.[15] Indeed, the lines of blame for Native and non-Native crimes had blurred noticeably, so that it was difficult to make any general conclusions about who was guilty of the violence. That same year, one of the most violent raids targeted two families, the Masons and the Camerons of Jack County. Evidence at the scene pointed to Indians: moccasined footprints, arrows in the walls and on the ground, unshod horse prints. No doubt regional citizens would have blamed the attack on hostile Natives as well but for the testimony of Cameron's eight-year-old daughter, who lived through the vicious assault. She claimed a white man with red hair had approached their house, and when her father went out to greet him, the stranger had murdered her papa. The only reason the girl lived was that she hid in a trunk while the house was pillaged. Later, the *Dallas Herald* ran an article stating that reservation Indians had spotted some of the Cameron family's personal effects—and a red-headed white man—in a Kickapoo camp.[16]

The reservation Indians, who were not immune from these non-Indian criminals either, became the scapegoat for frontier livestock thefts.[17] John R. Baylor's outspoken criticisms of Indian agent Robert Neighbors in the *White Man* drew the ire of those who made a livelihood selling cattle to the reserves. Claiming some of the ranchers had threatened his life, Baylor had asked Captain Ford for a bodyguard to protect him and his family. In 1858, Baylor asked two of the men stationed at his house to help him pull off a ruse raid on one of his neighbors. Matthew Leeper, an Indian reserve agent, claimed the two men related to him that Baylor had said he possessed:

> An excellent bow, and a large quiver of arrows, precisely such as are used by the Comanches on the Reserve (for they were purchased from them). That he could use a bow as well as a Comanche, and if they would go with him, (in the night time) to the house of a Mr. Shaw, a neighbor not far off, they would destroy his bee-gums and crops, and he would shoot arrows into [Mr. Shaw's] hogs and cattle, make trails and drop arrows in the direction of the Agency, that it would be charged to the Comanches on the Reserve.[18]

Amazingly, Baylor did not deny the accusation, but claimed his statement was made in jest.[19] Also in 1858, soldiers from Camp Cooper discovered the body of a man they thought had been the victim of vigilantism, which had become more and more prevalent in the frontier. What made the occasion memorable was a note found on the body that verified the existence of a rustler ring and its method of staging crimes so that the blame would fall on the Indians. The letter stated: "tell our friend at the *White Man* above all things to keep up the Indian excitement."[20] Clearly, the letter incriminates non-Indian outlaws for at least some of the depredations attributed to the Comanches and Kiowas.

Baylor and his followers won the war of propaganda. Angry mobs, blaming the Reserve Indians for the killings and livestock thefts that plagued the frontier, forced the Native prairie dwellers out of Texas. Even after their removal to Indian Territory, the accusations followed them. Doubtless, outlaws continued framing Indians for their deeds. S. A. Blaine, the Indian agent for the Wichita Agency in Indian Territory, wrote to Sam Houston on April 23, 1860, addressing the frequent crimes attributed to his reservation's Indians: "We are still to be made the scapegoat upon whom is thrown all the crimes committed on your frontier whether the same be the work of the white or red man." He called on Governor Houston to form a committee to investigate the accusations, disclose a report to the inhabitants of the Texas frontier, and, moreover, clear the name of the Indians on the Wichita Agency.[21]

Unimpressed, or perhaps unaware of Blaine's assertion, Chief Justice J. P. Guinn sent a list of "Indian Depredations in Montague County" to Texas governor Sam Houston, supposedly verifying the great damage hostile Indians had done to his small section of the Texas frontier. Guinn noted the dates, victims, and livestock value in thefts local settlers ascribed to Indians. Of the 134 horses reported as stolen by Indians, eyewitnesses claimed to have actually seen Indians stealing or in later possession of only 17.[22] The victims blamed the other 117 horse thefts on Indians with purely circumstantial evidence, such as the sighting of moccasin tracks on the trail, arrows in dead livestock, trails leading to known Indian villages, or other "Indian signs."[23] If he saw the letter, Baylor must have smiled quietly to himself at the efficacy of journalism.

In times of drought, individuals can come up with inventive means of managing resources to turn a profit. The lack of water created a high demand for food items. While civilians struggled to obtain basic necessities, military personnel could rely on the government to deliver commodities to them. Access to cheap stores and the high demand outside the military for such items set up the perfect conditions for corruption. Colonel McCulloch learned of one situation in which Texas officers had purchased supplies from the commissary and were selling them at a profit to starving civilians. He wrote to Captain James B. Barry at Camp Colorado that any future attempts to profit in this manner would result in the "severest penalties of law."[24]

Once the Civil War started in earnest, citizens of Texas realized their unique situation. They lived in the only Confederate state with a sizeable hostile Indian population menacing its frontier, which in addition to obvious problems also led to paranoia over other internal threats. In the summer of 1862, these irrational fears came to a head in the northern and southern frontiers of Texas. The state's Conscription Act of April 1862 created opposition in the counties north of Dallas, which had opposed secession before the war and whose residents were for the most part not slaveholders. As those opposed to forced military service circulated a petition, some of the signers created a Union League intended to foment organized opposition to the draft and to act as a militia to defend their homes from Indian attacks. Fears that the Union League was colluding with Union troops and Jayhawkers led Texas Confederates to arrest more than 150 men, of whom they found forty-three guilty of treason and executed.[25]

Similar fears surrounded the Texas Germans in Hill Country, who had no stomach for sending their young men to die defending slavery. When a group of more than sixty draft-aged German men decided to make their way to Mexico to wait out the war, rumors began circulating that they were going to rendezvous with Union forces and spearhead an attack into Texas. Nearly a hundred Confederate troops tracked the party down to the banks of the Nueces River and attacked before dawn on August 9, 1862, killing nineteen would-be emigrants and capturing nine wounded men, whom they executed shortly after the battle. Texas was in no mood to tolerate anticonscription demonstrations in any form.

In 1863 a new wave of fears swept the frontier state, but without such gruesome results. In late August Samuel Roberts, commander of the northern subdistrict of Texas, came upon information of a plot "originating with disaffected white men" and negroes, who supposedly planned to kill all whites—man, woman, and child—excluding known abolitionists and "reserving only young women for wives of the blacks."[26] The reported plot gained little traction, presumably because the hangings the previous year had quieted rebellious citizens. Another scare occurred in the winter of 1863. This time General Bankhead Magruder of the Confederate army told Governor Murrah that sources had informed him of a conspiracy to release the inmates from Huntsville prison, who would then join the growing number of deserters congregating in northern Texas to support a northern invasion of Texas.[27] Although the plot seems hardly believable, it is evident that the tension in Texas, resulting from its stressful geographic location, kept many of its citizens on edge.

To address such concerns, Governor Pendleton Murrah and the state legislature made provisions for defending Texas with home guard troops. The state exempted sixteen frontier counties from providing soldiers for the Confederacy in the hope that men from these areas would volunteer for the Frontier Organization, created to defend the Texas hinterland. The dispensation attracted people avoiding conscription in such numbers that the legislature had to revisit the bill and insert the stipulation that only those citizens who actually resided in frontier counties prior to July 1863 were to be excused from Confederate military service.[28] Rather quickly, the Frontier Organization became more occupied with apprehending the deserters and draft dodgers congregating on the frontier than with protecting it from Indian raids or Union invasions.[29]

North Texas was especially prone to attracting large numbers of deserters. The region held a fair number of Unionists who had opposed the war in the beginning, and the Texas commissioners sent to negotiate treaties with the Five Tribes of Indian Territory had observed a mass exodus from the area on the eve of the Civil War: "From North Fork to Red River we met over 120 wagons, movers from Texas to Kansas and other free States. These people are from Grayton [Grayson], Collin, and Denton, a country beautiful in appearance, rich in soil, genial in climate and inferior to none in its capacity for the production of cereals and stock. In disguise we conversed with them. They had proposed by the ballot box to abolitionize at least that portion of the State. Failing in this, we suppose at least 500 voters have returned whence they came."[30] This outmigration was followed by a larger immigration of deserters, draft evaders, and refugees attracted to the same quality the commissioners had praised in the region: the ability to produce food. As early as 1863, Brigadier General Henry McCulloch informed his superior that more than one thousand draft dodgers were living in the wooded regions near Bonham, Texas. They gathered for safety in large bands of two hundred to four hundred men and maintained a vigilant system of pickets to inform them of the approach of any party. He also claimed they had "sympathizers all through this country."[31]

By 1864 the problem of desertion had become epidemic in Texas. The rugged conditions of serving in Indian Territory or Arkansas with insufficient supplies led many young men to run for the frontier area below the Red River, where authorities would have difficulty apprehending them. A common ploy among Texas troops was to claim their term of service had expired and leave before they could be proven otherwise. In response, the Confederate command in Texas issued General Order Number 7, which stated: "all officers are hereby directed to report the names of men so leaving to commanders, who will send cavalry to pursue and arrest them, and bring them back to their commanders as deserters."[32] The official term of enlistment for the troops covered by the order was from August 6, 1864, to

February 6, 1865, and the general order was issued January 12, 1865, leaving no doubt about the issue of desertion.

Still, it was difficult for authorities to distinguish between men evading conscription, military fugitives, and criminals, as large bands of men with various backgrounds began to congregate near the frontier in 1864. James Bourland, commander of the Border Regiment guarding the Texas frontier, reported the presence of such bands north of Denton, near the Wichita Mountains in Indian Territory, near the saltworks of Shackelford County, north of Fort Phantom Hill in Jones County, and in the Concho Valley south of Fort Chadbourne.[33] Note that all of the locations were close to permanent sources of water, as the military had placed the old forts strategically near creeks and rivers to obtain the life-giving fluid and keep their enemies from using it. The Shackelford County locale controlled a vital commodity in the nearby saltworks as well. An undercover agent reported that H. J. Thompson, captain of the First Frontier District stationed at Buffalo Gap in southeastern Clay County, planned to abandon his unit, drive a large herd of cattle west to the Big Wichita River, join a large band of deserters, and head for Kansas.[34] During the multiple crises of drought, war, and increased criminal activity, even established individuals were tempted to emigrate out of the region.

In mid-March, Henry McCulloch reported that a number of disaffected people were gathering along the Concho River, intending to move their families and stock to California, and that a large number of deserters were "flocking to them."[35] Captain Henry Fossett of the Frontier Regiment, sent to investigate the report, located the band and reported to Colonel James "Buck" Barry that its numbers neared five hundred. The group of families and deserters must have altered its destination, for instead of moving in a westerly direction toward California, they were moving north in a forty-strong wagon train, herding a large number of cattle on a road made by previous bands of deserters.[36] Another report mentioned sixty deserters pushing one thousand head of cattle towards the San Saba River.[37] Any group of men out on the lam needed beef to survive, for its nutritional benefits as well as for its cash value. The large numbers of bovines discussed in the reports indicate that cattle rustling was a well-utilized means for feeding these bands, as well as providing economic stability for those driving cattle north to the markets of the Union army and Kansas Indian reserves.

Due to the withdrawal of troops from Texas, the relocation of the Texas Indian reserves, and the closure of the Confederate market with the fall of

Vicksburg in 1863, legal markets no longer existed for Texas frontier beef. Thus, the famed Chisholm Trail and other post–Civil War cattle trails may have their origins in this illicit rustling and herding of cattle to Kansas markets, where the beef was purchased to feed Union troops and refugee Indians during the war and after. Local citizens and military officers became extremely concerned about these roving groups of cattlemen. James Throckmorton, another officer in the Frontier Regiment, defended the maintenance of his unit by claiming that disbanding the military unit would unleash the full furies of hostile Indians and Jayhawkers and allow the whole region to be overrun by "deserters, draft dodgers, and traitors."[38] A troublesome element of society was moving to the frontier in growing numbers, attracted by its lack of law enforcement and the economic opportunities and food from the plentiful cattle. Yet these undesirables were not the only threat to law and order on the frontier.

In some cases, the frontier militia was every bit as unscrupulous as were the much-despised deserters. In the competition for resources in a deprived environment, guns equaled power, and the temptation to abuse authority led to a new source of savagery. The ubiquitous John R. Baylor, who had recently been involved in the invasion of New Mexico and the subsequent Confederate defeat at Glorietta Pass, is a textbook example of that abuse of power. Upon his return to Texas, he obtained command of a Ranger unit ordered to guard the frontier, but it seems that the residents of this region needed protection from him and his men instead. William Quayle, commander of the First Frontier Regiment, appealed to Governor Murrah on behalf of the residents of Weatherford, Texas, in a missive mailed on December 27, 1863, from Fort Worth. In this letter, he asked for the governor's intercession because Baylor and his men had hung two citizens in their town, robbed some of the stores, and had stolen individuals' horses, all with only forty or so men under his command.[39]

Similar atrocities had prompted many to distrust all Ranger units and to rely instead only on Confederate troops. The families who remained on the frontier found the violence tripled from what it had been prior to the war. There was still the occasional Indian raid, but more pressing was the non-Indian violence,which caused so many who continued to live in the region to gather in multi-family stockades for protection. One such group took up residence near Camp Davis in Gillespie County, which housed Company A of the Texas Confederate military. Upon learning that this unit had been ordered to relocate to Fort Duncan, the citizens of Gillespie,

Kerr, and Kimble counties sent a petition to military authorities requesting that the troops remain at Camp Davis. They did not even mention Indian attacks, but instead pointed to the marauding of "jayhawkers and disloyal men who infest our frontier, burning our houses and murdering the good citizens." They claimed it had become so dangerous to gather in their stock that they only rode out in groups of five or six and still felt undermanned. These hardy settlers made it well known that they had no confidence in the capacity of the minute men, who could themselves be a cause for fear, to protect them, but trusted only Confederate soldiers to offer some source of safety.[40]

Conditions along the southern Texas frontier were no better. Area ranchers complained to military officials about their loss of cattle to "lawless bands of white men" and Indians who drove them across the Rio Grande. One cattleman reported the theft of more than five thousand head from the Fort Clark region.[41] Another area settler made his case more forcefully: "Indian depredations . . . are now of frequent occurrence, but still more dangerous and destructive [are the] depredations of deserters, jayhawkers, and robbers who infest the whole country from the Colorado to the Rio Grande. Without some force to protect this frontier I have no doubt the whole country west of San Antonio will be deserted by every loyal citizen."[42] One region's appeal for troops was followed by another's plea to leave the units in place. Obviously, there were not enough dependable companies to go around. Indeed, the whole frontier of Texas was "infested with a great number of renegades, bushwackers and deserters" who were waylaying unsuspecting travelers, taking men from their homes in the middle of the night, hanging them, and then robbing their homes as terrorized wives and children looked on. In the words of Brigadier General G. D. McAdoo, "the Indians seemed to be the least talked of, the least thought of, and the least dreaded of all the evils that threatened and afflicted the frontier."[43] There is no doubt that the region was extremely unstable due to the presence of so many militarized groups acting on both sides of the law.

By 1865 the Texas frontier and its people were absolutely exhausted. Many families had lost loved ones in the war, while others had relatives who deserted to escape the horror of battle and the deprived conditions of camp life. Texas soldiers reading letters from home would hear of their families' destitution resulting from the war and the drought, which continued through the last year of the conflict. As news reached the state of Lee's surrender in late April, the soldiers knew the war, for all intents and purposes, was over, but Confederate diehard General Kirby Smith still refused to surrender, and a sizeable force of Texas soldiers near Brownsville were of a like mind. When Union officers ordered the occupation of Brownsville at the southern tip of the Lone Star State by a five hundred–man force, half African American soldiers, the result should have been predictable. The last battle of the Civil War took place on Texas soil on May 13, 1865. Confederate troops granted no quarter to the men of the Sixty-Second U.S. Colored Infantry in a battle that should never have been fought. Ironically, the Battle of Palmetto Ranch, as it became known, was a Confederate victory, but of course it did nothing to change the outcome of the Civil War.

The next day, Union and Confederate officers negotiated a truce, and the Texas troops simply melted away, heading home with their rifles. On the way, some broke into stores, taking flour, beans, and other provisions to ease the suffering of their families back home. A group of fifty men brazenly robbed the Texas state treasury, relieving the vault of $17,000 worth of specie. One wonders how these men explained the money and supplies when they reached their destinations, and certainly, some wives simply didn't ask. They were relieved to have their husbands back home, especially if they brought some food.

The drought had prevailed in the region for a decade, reducing the options for its inhabitants to obtain sustenance. Returning war veterans and area settlers faced the difficulty of procuring a livelihood in a parched environment. Farming without irrigation failed to produce crops worth harvesting, and the withdrawal of the U.S. military installations and the Indian reserves removed the demand for beef. The closest cattle market was in Kansas, but the war made driving the animals there a dangerous venture, one that only those living outside of the law cared to exploit. The lack of cattle markets explains why so many people in the region were willing to blame all of the crimes on Indians: if they could convince the state or federal government of the necessity to station troops on the frontier again, the markets for their livestock would return.

Without such avenues for making money and with the increased competition for resources like cattle and horses, many frontier families moved back east. Major George Forsyth thought the reports of Indian depredations were indeed exaggerations, intended to incite the U.S. government to send the military back to the Texas frontier, bringing with it needed economic opportunity. In 1866 General Philip Sheridan ordered Forsyth to investigate the conspicuously high number of crimes attributed to Indians on the Texas frontier. Forsyth found no justification for the accusations. In his report, he claimed: "Thus far I have not been able to get any accurate information in regard to the Indian Depredations, but I am convinced that many people who are moving from the frontier are doing it to better their condition and not from fear they may have of the Indians; for instead of stopping in the first well settled and perfectly safe counties they come to, they are moving down towards the central portion of the State, within reaching distance of the railroads so that they may be sure of a market for what they produce."[44]

By fortuitous coincidence, as the Confederacy accepted military defeat, the decade-long dry trend came to an end. The people who survived a war that threatened the nation's integrity, an economic depression in the Panic of 1857, and an environmental crisis more severe than that of the 1930s drought might compete for the moniker of America's Greatest Generation because through all that, they somehow kept the nation intact. The rains returned, the bison followed the water and grasses, and settlement of the frontier continued as if the mid-century catastrophes were so much spilled milk. After the war, another generation of settlers would fix their land-hungry gaze on the southern plains and the tallgrass prairie, this time at the encouragement of railroad companies and real estate developers.

It is a basic tenet of mankind's interaction with the environment that nomadic people perfected their lifestyle when the landscape and weather made it difficult to create permanent villages and homes. The geography and climate of the southern plains had for eons encouraged constant move-ment by animal and human populations, but the situation was soon to change. By the mid-1870s, the U.S. government had successfully removed the Native peoples of the region to reservations in Indian Territory, open-ing up the wide expanse of prairie to non-Indian settlement. The grasslands were a region of undependable rainfall, but it had provided grazing for bo-vines and sustained Natives who followed the herds in their travels for cen-turies. The nomadic people had been forced onto stationary reservations, their culture of hunting and bartering supplanted by a sedentary culture in a region more suited to to their old ways of extensive land use, avoiding intensive agriculture and livestock management. The new inhabitants took little notice of such concerns, but the climate and geography remained the same. What other result could there be, but disaster?

ELEVEN

Boomer Bust

The fabled American homesteading dream broke down on the devil's anvil of the plains.

JOHN OPIE

It is amazing how quickly the human mind forgets. Americans tend to exhibit selective memory about our own history, especially when environmental perceptions conflict with the promise of economic advancement. Many plains settlers believed that there could never be another drought like that experienced during the early 1860s.[1] Moreover, boomer literature stressed ample rainfall, healthy living conditions, and good soils on the southern plains. The post–Civil War era of promotional literature spurred the initial period of non-Indian settlement of the southern plains and tallgrass prairie in the late nineteenth century, a time that was for the most part more humid than normal.

Of course, there were a myriad of reasons why non-Indian settlers began swarming over the southern grasslands in the later 1800s. To begin with, promoting plains settlement made good politics, while the defeat and relocation of the southern plains tribes from the sprawling grasslands to Indian Territory reservations opened up the grasslands to non-Indian settlement, and the extension of the rail system promoted land sales. Also, as Walter Webb so aptly demonstrated in his seminal work *The Great Plains,* the technological advances of the time simplified the business of agriculture, the steel plow cutting through the dense root systems of the plains' native grasses.[2]

Various entities joined in the effort to populate the plains with non-Indian settlers. Even prior to the Civil War, movement west had become a political platform. The debate over the extension of slavery triggered Free Soiler advocacy of a homestead law. Nothing could thwart the spread of slavery like thousands of 160-acre farms, which were too small to support the "peculiar institution." Abraham Lincoln and the Republican Party also favored increased immigration and a homestead act as a means of drawing thousands of people from Europe to the United States and, one might add, to the Union armed forces. As soon as the culmination of the Civil War allowed the nation to return its gaze to frontier development, these groups began to call for resuming settlement of the region. The Republican Party looked to the inhabitation of the nation's interior as a way to refocus the energies of the devastated country and prove to other nations that the United States was still vigorous and growing.[3] A U.S. senatorial group traveled to the vicinity of the one hundredth meridian near Fort Harker, Kansas, in 1867 to evaluate the region's possibilities. This was a wet year, with playas full of water and the verdant grasslands teeming with wildlife. It should come as no surprise that one of its members, Illinois Senator Lyman Trumball, concluded that the "American Desert" label was a mistake.[4]

Other government officials were also convinced of the blossoming of the plains. During the late 1860s and early 1870s, Ferdinand Vandiveer Hayden forcefully promoted the theory that rainfall was increasing on the plains. Born in Massachusetts and a graduate of Oberlin College in Ohio, Hayden earned his medical degree in 1853 and soon after joined an exploratory expedition to Nebraska Territory. While the slavery debate intensified in his native state, Hayden was out West collecting floral specimens and gathering data for a report on Yellowstone country and the Missouri River. During the war, he served as a physician, rising to the rank of chief military surgeon for the Union Army of the Shenandoah. After the war in 1869, he was back in the West as the head of the United States Geological Survey (USGS). In this capacity, he proved himself a promoter of business interests in the West and tirelessly supported theories that rainfall could increase in the area once known as the Great American Desert. In his "First Report of the United States Geological Survey of the Territories, Embracing Nebraska," Hayden claimed that settlers in Nebraska had planted trees to produce shade in a hot summertime environment, but that an added benefit of this endeavor was an increase in rainfall in the state.[5] He echoed a faith in afforestation that had earlier roots.

The "Gospel of Tree Planting" was one of the first theories dealing with increasing rainfall on the southern plains. After the drought of the 1860s, farmers of the Midwest sought methods for improving the moisture retention of the soil. Many believed trees could help this process by providing windbreaks, lessening the impact of wind on evaporation. By 1868 eastern counties of Kansas were home to substantial groves of imported Austrian and Scotch pine, honey locust, elm, eastern red cedar, and Osage orange trees, funded by a state program that offered cash payments to those who maintained a copse or line of trees for three years or more.[6]

Railroad companies joined in encouraging the afforestation of the southern plains. The U.S. government had subsidized the construction of rail lines across the West by offering land grants to the companies to help defray the cost of construction. The railroad companies knew they would benefit from the settlement of the prairies and plains through the sale of the subsidized land. Larger populations along their tracks would translate into increased freighting of products on their lines, which would be a great boon to their enterprises. The Gospel of Tree Planting offered a means of recruiting settlers by providing hope that people could manipulate their environment to increase rainfall on the plains.

The Kansas Pacific Railway was at the forefront of research in afforestation. In 1870 its managers appointed Richard Smith Elliott as industrial agent and financed the construction of three agricultural experiment stations on the western plains of Kansas to test Hayden's pronouncement. It was the railroad's hope and Elliott's conviction that these stations would prove that rainfall was increasing, so the agent supervised the planting of barley, wheat, rye, and corn at all three stations and constantly checked on their progress. His reports pointed to increased moisture and yields and were far from being dismissed as quackery: one of his articles graced the pages of the *Annual Report* of the Smithsonian Institute.[7] However, the drought and economic panic of 1873 appeared to rebut his theory and injured the Kansas Pacific's financial resources. The company terminated the experiment stations that same year. Still, Elliott's influence on booster literature was evident up until the 1890s.[8]

Despite the setback, tree planting spread on a national level as well. The states of Minnesota and Wisconsin had already adopted afforestation measures to allay the effects of the lumber industry's massive clear cuts in 1867 and 1868. J. Sterling Morton, an influential newspaper editor who had served as secretary of Nebraska Territory, suggested to the Nebraska State

Board of Agriculture in 1872 that it promote a celebration of tree planting by designating a single day in which all citizens were encouraged to plant trees across the state. The legislature could furthermore give a cash award to the individual who planted the most trees. The board accepted Sterling's proposal and initiated Arbor Day that year. The concept was so popular that the state legislature proclaimed it a state holiday in 1874.

The U.S. legislature officially recognized the importance of planting trees when Congress passed the Timber Culture Act in 1873. The law gave 160 acres of public domain to any head of household who would plant forty acres of timber and maintain it for ten years. Within five years of its passage, congressmen amended the bill, easing the requirement to ten acres of timber to be maintained for five years in order to encourage more homesteaders to take advantage of the law.[9]

The Gospel of Tree Planting took on new significance as politicians and members of the scientific community supported the theory. In support of the Timber Culture Bill and with hints of belief in human-induced climate alteration, Phineas Hitchcock, Republican senator from Nebraska, stated that "the object of this bill is to encourage timber, not merely for the benefit of the soil, not merely for the value of the timber itself, but for its influence on climate."[10] In 1883 the head of the nation's Forestry Division of the Department of Agriculture, Nathanial Egleston, also promoted the concept, claiming that trees could have a "direct influence . . . on the distribution of rainfall."[11] Such a sweeping statement could not go unchallenged for long. In 1886 Dr. Bernhard Fernow crossed many promoters of sylviculture by claiming he doubted the influence of trees on rainfall. One can imagine the disappointment of tree-planting faithfuls when they heard of Fernow's appointment to direct the Forestry Division, but Fernow made up for this negative perception by urging the establishment of research programs at state colleges.[12] By 1890 seventeen colleges had some form of forestry curriculum.[13]

The Kansas legislature also actively promoted the settlement of its state. In March 1867 the state congress established a Bureau of Immigration with the intent of "filling the blank space heretofore allotted to the Great American Desert."[14] It seems the ghost of the Long Expedition still lurked forty-seven years later in the perceived geography of the plains. Kansas legislators attempted to overcome this obstacle by using state agricultural agencies to promote settlement. In 1877 the state legislature passed a bill forming a board of agriculture, which became a vigorous promoter of husbandry, in the hope that they might increase the population of Kansas by attracting more farm families to the state.[15]

However, the most successful recruiters for settlement of the Kansas plains were the aforementioned railroad companies, who were eager to overturn the image of the "Great American Desert" to increase sales of their company real estate.[16] They claimed that the southern plains, rather than being limited by aridity, were in reality marked by incredibly fertile soils, ample rainfall, and a healthy atmosphere.[17]

The best example of these recruiting efforts can be found in the activities of the Atchison, Topeka and Santa Fe Railway (AT&SF). The U.S. government had granted a whopping 3,000,000 acres of land along the railroad's right-of-way directly to the AT&SF. Company directors quickly established the rail lines to more quickly recruit settlers. They hired A. E. Touzalin to direct the land promotion department from a central office in Topeka, Kansas. Touzalin proved extremely creative in organizing the operation. He initiated a newspaper advertisement campaign and hired a small army of staff personnel, including clerks, correspondents, land agents, newspaper reporters, and advertising solicitors, to assist him in boosting the region. Touzalin devoted one whole department of his agency to foreign immigration. The director of this division, C. B. Schmidt, traveled to Europe, recruiting settlers in person and going so far as to cross into Russia illegally to visit German-speaking Mennonite farming communities there in February 1875. While on his journey, Schmidt appointed local European businessmen to act as Santa Fe agents in Germany, Austria, and Switzerland.[18] Schmidt's sales techniques must have been very convincing, for later that year, 1,900 Mennonites migrated from Russia to Marion, McPhearson, Harvey, and Reno counties in south-central Kansas.[19] These hardy settlers will forever be lauded by American agriculturalists for their importation of Turkey Red, a winter wheat that grows well in the semi-arid climate of the plains, but a stowaway aboard their sacks of grain was the tenacious Russian thistle. This plant grows into a clumpy shrub from spring to fall, then breaks from its roots in the winter to tumble wherever the wind blows it. One plant can contain a thousand seeds, and each time it hits the ground as it rolls across the landscape, it disperses them. In the United States, the Russian thistle is called tumbleweed, and modern plains drivers curse it as they swerve to avoid hitting the wind-driven plant skeletons.

To the south, Texas boosterism and settlement followed a similar pattern, but before any real estate developers could sing the praises of plains settlement, the Native occupants had to be convinced to give up their ways of life and accept confinement to reservations elsewhere. After the return of the bison to the southern plains and as the Civil War and mid-century drought passed into the historic lore of the region, non-Indian buffalo hunters began combing the grasslands, locating sizeable herds of the large ungulates. Unlike the Natives, these men did not need the buffalo's meat. All they took were the tongues and hides, which could be used for machine belts in the growing number of factories back east. This specialized form of hunting left thousands of carcasses to litter the plains, inadvertently and unexpectedly boosting the regional wolf population. As the bison numbers plummeted from overhunting, the plains tribes found themselves without a much-needed food source and a major trade commodity, buffalo hides, to barter for food products and weapons, surely impairing the tribes' fighting capacity. In 1875 Colonel Ranald Mackenzie, commanding the Fourth U.S. Cavalry, defeated the Comanches at the Battle of Palo Duro Canyon in the Texas Panhandle, removing those who still attempted to maintain their traditional way of life. The destruction of their homes and horses forced large numbers of the Comanche tribe to walk to Fort Sill in Indian Territory and surrender, as winter was fast approaching.

The defeat of the Comanche opened up the Texas Panhandle to non-Indian settlement and a wave of settlers arrived the very next year. Hispanic sheepherders from New Mexico moved in from the west as Anglo cattlemen migrated from the east and north. A new type of cattle industry had grown out of several post–Civil War developments: the growing urban population in the nation's northeastern section served as a huge market for beef products; the introduction of the refrigerated railroad car in the 1880s allowed meat to be transported long distances from the nation's meat packing center in Chicago; and the construction of rail lines into the southern plains provided a fairly close shipping point for most cattlemen. Ranching activities generated sizable profits and attracted outside investment, especially from Britain. The abundance of unoccupied land and increasing demand for beef led to a bonanza on the southern plains that required little boosting.

In Texas, the state, not the federal government, retained the rights to public lands within its borders. As a result of Texas's prior status as an independent republic, the United States had agreed to that stipulation in the annexation treaty with Texas. Aware of the liberties railroad promoters took in exaggerating the climatic and soil qualities of the southern plains in order to boost their land sales, the Texas state legislature decided to grant land to rail companies that was not adjacent to their lines and to give them only twelve years to dispose of the property before it reverted back to state ownership, thus pressuring the railroads to sell the land quickly.[20] Still, as in Kansas, the railroads were the primary agents in promoting the settlement of the Texas Panhandle.[21]

Unlike the eastern portion of the state, in the high plains the railroads arrived prior to widespread settlement. Although the town of Clarendon existed as a Protestant Christian colony in 1878, there were few farms or communities in the region. Instead, cattlemen operated under an open-range system that utilized the state lands as one giant commons for any-one's use. Thus, the transport companies had a different set of goals for the southern plains than did the area's ranchers, a difference that would be exaggerated when the rail lines reached the vast grasslands. The Fort Worth and Denver Railway reached Clarendon in 1887; the Atchison, Topeka and Santa Fe and the Rock Island came to the Panhandle shortly thereafter.[22] Similar to the company's strategies in Kansas, the AT&SF railroad placed $50,000 annually in their advertising budget to employ staff writers and produce newspaper advertisements, pamphlets, tracts, and monthly mag-azines, all aimed at recruiting settlers.[23] Other businessmen used similar strategies. Local and state organizations, newspapermen, real estate agents, and speculators all joined the rail companies in promoting the southern plains for settlement, and their efforts were quite successful. A report of the Texas Bureau of Immigration stated that 125,000 people immigrated to the state in 1873. In 1875 that number had risen to 300,000, and the high-water mark was reached the next year, with 400,000 new home seekers arriving in west Texas alone.[24]

Between Texas and Kansas lay Indian Territory, which was unique in that it was spared widespread non-Indian settlement during the 1870s and for most of the 1880s. As states to the north and south began filling up with settlers, Indian Territory remained a reserve for the Five Tribes, the south-ern plains tribes, and other refugee Natives. The Civil War had brought physical, emotional, and political catastrophe to the residents of Indian Territory, regardless of which side they fought for. Sadly, even though many Cherokees, Muscogees, and Seminoles had fought for the Union, the fact that all of the Five Tribes had signed peace treaties with the Confederacy gave U.S. government officials the pretense to reduce tribal holdings.

In 1866, shortly after the war, the United States negotiated new treaties with the Five Tribes, and all Indian governments lost control of their western lands. The lone exception was the Cherokee Outlet, and even portions of this tract were used to resettle the Osage, Tonkawa, Ponca, and Otoe tribes from Kansas and Texas. The Choctaws and Chickasaws forfeited their claims to land west of the ninety-eighth meridian for $300,000.00. The Muscogees relinquished the western half of their reserve for $0.30 an acre, while Seminole leaders signed a treaty giving up all of their holdings in Indian Territory for $0.15 an acre, then were cajoled into purchasing land from the Muscogees at $0.30 an acre. This series of land transactions left a sizeable region right in the center of the territory not claimed by any entity. This area was aptly labeled the "Unassigned Lands," and everyone knew that it was just a matter of time until it was opened to non-Indian settlement.

After the Medicine Lodge treaties in 1867, the Kiowas, Comanches, and Kiowa Apaches (KCA) relocated to the Wichita Agency, a reserve taken from the old Choctaw and Chickasaw Leased District to the south of the relocated Texas reserve Indians. The limits of the KCA reserve were bounded by the Red River on the south, the ninety-eighth meridian on the east, and the one hundredth meridian in the west. The Cheyennes and Arapahos also agreed to settle immediately to the north of the KCA reserve and the Wichita Agency. Of course, the Caddos and Wichitas had a background in agriculture, but the Kiowas, Comanches, Apaches, Cheyennes, and Arapahos did not have this knowledge or societal tradition to rely on. Making successful farmers out of these expert hunters would be difficult.

Precipitation levels remain absolutely unpredictable on the plains, but especially during the post–Civil War era from 1870 to 1880, rainfall fluctuated between low, average, and high and created agricultural problems in the region. The unreliability of vital moisture for crops made farming a gamble every year. Indian agents worked with each tribe, supervising their farming activities and reporting on their successes and failures to the regional commissioner, who prepared a yearly report to the commissioner of Indian affairs. Their communications document various tribes' struggles to coax crops from drought-stricken soils during these years. In 1874 drought and the reduction of ground cover by numerous prairie fires destroyed the corn crop at the Wichita Agency.[25] In 1879 John Shorb reported that the Sac and Fox opted to deal with dry conditions as nomadic people had done for centuries: by relocating, or in this case, by driving their livestock "a great distance from their reservation, where they found water and grazing ground, most of them remaining the entire season."[26] The drought continued up through August the following year, as dry and parched grass dominated the territory and frequent prairie fires swept across the highly combustible cover, according to Ponca agent William Whiting.[27]

Similar reports came from supervisors of agricultural tribes, people who had been self-sufficient through agriculture for generations and who had proven that during wetter years, their yields were more than sufficient. From 1875 through 1878 the Wichitas were able to gather a surplus of crops, bringing in bountiful harvests of corn, wheat, and oats.[28] These were also bumper years for the Sac and Fox farmers who, as their agent reported, had "abundant crops" in 1877 and "a surplus of corn" in 1878.[29] However, one can imagine how the nomadic tribes fared with their experiment in agriculture during the dry years.

Even as the unpredictability of rainfall made farming a treacherous enterprise, so-called Boomers began advocating for the opening of the Unassigned Lands to non-Indian settlement. As Kansas and Texas experienced incredible immigration, the number of available attractive homesteads in those states diminished. Yet imaginative booster advertisements continued to lure settlers to these areas. With new arrivals finding that they had arrived in the region too late to claim choice plots, it was not hard for those calling for the opening of the Unassigned Lands to gain followers. By 1879 there was a sizable movement arguing that the Unassigned Lands were public domain and therefore open to settlement under the Homestead Act. At the forefront of the campaign to open these lands stood the railroads. Two of the principal publicists who advocated for the opening were Judge T. C. Sears and E. C. Boudinot. The Missouri-Kansas-Texas Railway engaged Sears as a company attorney, and some believe that Boudinot was attached to the same firm.[30]

Perhaps the most notorious individual connected with the Boomer movement was David L. Payne. Although a household name with a county namesake in modern Oklahoma, it appears that Payne was not much of a success at anything. He had been fired from a postmaster position in Leavenworth, Kansas, for irregular accounts, and he had already lost two of his three titles to quarter-sections in Kansas due to an over-extension of finances.[31] Furthermore, he was a "chronic borrower" who was rarely seen at tasks requiring hard labor, even though he supported a six-foot, four-inch body that carried all of two hundred fifty pounds.[32] The hapless Payne lived

David L. Payne's Last Boomer Camp, March 1883. Photograph by Carl P. Wickmiller. Thomas N. Athey Collection no. 4985. Courtesy of the Research Division of the Oklahoma Historical Society.

with, but never married, "Ma" Haines, who bore him a son they named George. In 1879 the ne'er-do-well finally found a mission that stimulated the sizeable ego to match his robust frame. Payne formed an organization known as the Oklahoma Colony and sold 5,000 shares to the St. Louis–San Francisco Railway.[33] Finding the topic of opening Indian Territory lands to be as emotionally charged as it was lucrative, he sought the support of promoters from Wichita, Kansas, and established another of his Boomer organizations, the Oklahoma Town Company.[34]

Between 1880 and 1884, Payne and his followers attempted eight illegal entries into Indian Territory. The U.S. military captured him on all of these forays and either escorted him out or held him in confinement, pending trials. After only his second invasion, soldiers escorted Payne to Fort Smith to stand trial for trespassing on tribal lands and government property. While on the journey to Arkansas, the officer in charge of the military escort claimed that Payne had complete confidence that he would emerge from the trial unscathed because he had the best legal advice in the country and had followed his instructions explicitly. This bravado led the officer to the conclusion that Payne and his followers were under the guidance of the major railroads.[35] At Fort Smith, Judge Isaac Parker found him guilty of trespassing on Indian lands. The judge fined Payne $1,000, but because he had no property or money, officials were never able to collect the penalty.[36] On another attempted filibuster and subsequent capture, the Boomer leader, obviously drunk, threatened to cut the throat of the first soldier to lay hands on him.[37] On November 27, 1884, while partaking in a Thanksgiving Day breakfast at the De Bernard hotel in Wellington, Kansas, the larger-than-life Payne suffered a massive heart attack and perished. After his death, the reins of the Boomer movement passed to W. L. Couch, who pursued its goals with as much vigor as the fallen leader.

The efforts of the Boomers came to fruition in early 1889, when Secretary of the Interior William Vilas negotiated the purchase of the Unassigned Lands from the Muscogees and Seminoles, and Congress authorized President Benjamin Harrison to open the area to settlement. Harrison wasted no time in opening of the Unassigned Lands, proclaiming them available for settlement on April 22 and initiating the first of a series of land runs in Indian Territory, out of which the "Oklahoma District" would grow.

In the face of all this boosterism, the southern plains experienced a severe drought from late 1885 to early 1887. While as a rule this region usually receives low amounts of rainfall during the winter months, the late fall and winter of 1885–1886 were more dry than expected. Sandstorms struck Fort Sill, Indian Territory, and were reported as far away as Fort Union, New Mexico, and Cleburne, Texas, during October and December.[38] From November to February, Abilene, Texas, saw no single month register over one inch of precipitation.[39] The drought that followed for the next fifteen months was most severe in a narrow triangular region running from Weatherford, Texas, to Dallas, Texas, and northwestward to Fort Sill, Indian Territory, where rainfall statistics reveal collections of only 15, 28, and 24 percent of those respective towns' average precipitation for the winter months.[40]

As spring of 1886 came to a close, the environment exhibited the telltale signs of drought distress. In May the usually reliable Colorado River went dry near Brady in McCulloch County, Texas. The town of Cisco in Eastland County, Texas, was so desperate that the Houston and Texas Central Railway ran a daily car load of water out from Albany and sold the precious liquid by the bucket or barrel.[41] Local ponds and creeks dried up into pools so small that it was easy to catch fish by hand, but difficult to locate water for livestock. Many ranchers had to move their stock, some as far as thirty miles, to find a suitable location to slake their animals' thirst. They camped out near their herds, hoping to outlast the drought in this migratory manner.[42]

Ranchers in Baylor, Throckmorton, Archer, and Young counties of Texas were so desperate to find grass and water for their cattle that they simply turned the animals loose, allowing them to follow the dry creek and riverbeds downstream. The cowhands did not try to control the frantic stock, but let them destroy cultivated fields and consume the water that remained in the ponds and water holes of the farmers down river. As of late July, thirty thousand head of starved and thirsting stock had encroached on the water sources and fields of Jack and Wise counties. Farmers in the threatened areas organized to oppose the invasion of these hooved locusts. A violent collision between the cattlemen of the migrating herds and the ranchers and farmers of Jack and Wise counties was averted only by a timely rainshower to the west. The two groups held a meeting, and the cattlemen agreed to move their stock back west where the recent rain promised more water.[43]

Elsewhere in the drought-plagued region, cattle perished by the thousands.[44] In New Mexico, the Pecos River stopped flowing above the confluence of Salt Creek. Cattle gathered at the dwindling water holes, churned them into mud puddles, and expired in the the heat with their tongues swollen black.[45] Ranchers unloaded their herds on the cattle market, causing prices to plummet; steers that had brought up to forty dollars in 1885 were worth only eight dollars in 1886.[46]

Drought created another unforeseen problem for the region's farmers and ranchers. The low amount of water in the Red River made it shallower, and thus more susceptible to freezing. When the following winter proved to be extremely cold, the river indeed froze over from bank to bank, presenting a new danger to the livestock in the area. When cattle encounter a frozen water source, they attempt to stomp through the ice and slake their thirst near the shore. When this does not yield the desired result they move toward the middle of the water source where the ice is more thin, only to plunge to a frigid death when they break through the fragile barrier. Ponds are extremely susceptible to icing over in cold temperatures, but rivers rarely do because moving water is much more difficult to freeze. The freezing of the Red River meant that what had been an almost constant source of water during the winter was now unavailable to the cattle. Andy

Addington, who had a ranch along Bear Creek near the present town of Velma, Oklahoma, recalled, "I lost about four thousand head of cattle. . . . They died for want of water. All of the creeks went dry . . . up and down the Red River. . . . We dug wells and everything else trying to get water. Red River froze over. . . . It froze as a result of the drouth, and we had a very severe winter."[47]

One can imagine how hard the drought hit the area's farmers as well. By mid-July citizens of Albany, Hulltown, and Breckenridge, Texas, were circulating petitions to the governor asking for relief.[48] Within the immediate area, those with any means did their best to aid the destitute. People of Ballinger, Texas, though suffering from a lack of rainfall as well, sent a wagon full of flour, bacon, and meal to the devastated residents of Content, Texas, and followed this act of generosity with twelve other loads of supplies within the month. But the supplies they sent may not have been backed by altruistic motives: the town leaders of Ballinger were campaigning to move the county seat from Runnels City to their town and sorely needed the votes from Content to accomplish this.[49]

There were also private efforts to provide relief. East Texas counties held meetings to assist the drought victims, but these were not well attended and failed to draw much support. Still, Wilson County citizens sponsored a boxcar of provisions for dry Runnels County, Texas, and citizens of Weatherford, Texas, purchased ten thousand bushels of wheat for resale at cost to those in need in devastated Parker County. Likewise, in an act of purest optimism, *Farm and Ranch* magazine donated a railcar full of mixed planting seed to the relief effort with the belief that rains would resume the next year. In a more direct effort to ease the suffering of the drought victims, the *Fort Worth Gazette* and *Dallas News* cosponsored a drive that raised enough money to give every county in the drought-stricken area $750 with which to purchase famine relief. Other civic organizations and individuals sent boxes of clothing and canned goods as well.[50]

Larger donations arrived from outside the borders of Texas. The Merchants Exchange of Saint Louis, Missouri, sent ten carloads of provisions, and the Union Stockyards of Chicago provided $1,850 to Governor Ireland of Texas for dissemination to the needy.[51] Still, the suffering in the region was great and leading citizens of Albany, Texas, took up a collection to send local Reverend John Brown on a tour of northern cities to solicit aid. The reverend traveled to Chicago and spoke at the Produce Exchange, obtaining a $5,000 donation for his efforts.

Amidst all of this distress, there were those who opposed attempts to raise awareness of the southern plains drought in the nation's metropolitan centers. Many West Texas cattlemen hoped that the farms would fail so that their farmers would move back east, opening up the public domain for cattle. Ranchers required large landholdings to operate due to the scarcity of water and forage and therefore saw the region as fit only for large cattle spreads, not small farm plots. Another group that hoped to stall Brown's collection efforts was composed of local town boosters such as real estate developers, store owners, and newspaper editors who feared that the reverend's mission would dissuade people from moving to and settling in the western prairies. They thought the negative publicity would damage booster efforts. Rail companies also attempted to convince Brown to return to West Texas, but he refused and instead visited New York City, where he raised five hundred dollars from the New York Produce Exchange.[52] The railways paid and sent their own lobbyists to shadow Brown and refute his arguments.

Reverend Brown's trips to Chicago, New York and, later, Washington, D.C., did make many aware of the deprivation induced by the southern plains drought. In Congress, S. W. T. Lanham, a Democratic representative from Texas, proposed allocating $50,000 to purchase seed for the drought area. Before this proposal reached the floor of the Senate, agents working against relief were able to convince many senators that Brown had grossly exaggerated the dry spell. As a result, the senate reduced the proposal to $10,000, but congressional efforts to relieve the suffering of drought victims went for naught, as President Grover Cleveland vetoed the bill as unconstitutional. The president did give a $25 donation from his personal account as a gesture of Christian charity.[53]

All of these contributions began arriving in Texas by August 1886, but each county had to devise its own methods for disbursing the relief. In Taylor County, a judge organized a relief committee to oversee the dispensing of aid on the basis of need. Each school district in the county elected three men to list the most impoverished families in their district, and the relief committee referred to these lists to disburse funds.[54] On December 27, 1886, county judges from the water-starved region met to establish a system that would ensure an equitable distribution of aid. It is estimated that in the twenty-one counties represented in the meeting, there were 30,000 individuals living in complete destitution. The judges appealed to the Texas state legislature to provide at least $500,000 in relief, and they also decided to organize a national campaign to raise donations. This last proposal came under intense criticism. Inhabitants within the larger drought area who had received some rainfall did not think it was necessary to tarnish the image of West Texas by soliciting aid. Of course, those experiencing the driest conditions fully supported the proposal.[55]

In the meantime, 1886 was a gubernatorial election year in Texas, and the drought had created a topic for election platforms. In 1887 the new governor, Lawrence Sullivan Ross, supported a bill appropriating $100,000 for famine relief. A committee of three men traveled to the area of concern and assessed what portion of the money each county would receive, with which it could purchase corn, flour, and meal. Nearly 29,000 people from 37 counties in West Texas received an average of $3.25 of aid, with the highest percentage of recipients coming from Baylor, Shackelford, and Stephens Counties.[56]

After such a dry year and a half, the ground cover had begun to die out completely, exposing the earth to the vagaries of the wind. Eighteen eighty-seven was another year of dust storms. In January Midland, Texas, endured seven blackouts, and Abilene was hit six times. The episode of January 19th was the most intense: on that morning, the storm obscured the sun from view until two hours after sunrise, causing some, no doubt, to think the end of the world was near.[57] Through February, March, and April, dry conditions prevailed. Those who had managed to maintain their cattle herds during the preceding year were forced to sell at rock bottom prices or watch their animals perish from thirst. Many ranchers started leasing land from the Cherokees and other tribes of Indian Territory and shipping their livestock to the areas that were not yet overgrazed. Still, Indian Territory was not much better off. Fort Gibson suffered through three dust storms in late April alone.[58]

Not surprisingly, the drought was attended by outmigration. People began moving back east to find new farm plots, to obtain other work in order to keep their families fed, or to live with relatives until the drought broke. On the Llano Estacado in western Texas, 90 percent of the more than one thousand settlers forfeited their claims as a result of the drought.[59] Even as far east as Blanco County, near the modern town of Menard, one disgruntled farmer inscribed a message on his cabin floor that read:

250 miles to the nearest post office
100 miles to wood
20 miles to water

6 inches to hell

God Bless Our Home!

Gone to live with the wife's folks.[60]

Dr. Kindall, the editor of the *New York Evangelist,* visited the drought area in 1886 and claimed that he witnessed forty-five wagons moving east through Jacksboro in one day.[61] Scholars have estimated that nearly 50 percent of the Texas high plains population moved east during 1886 and 1887.[62] The census figures for 1880 and 1890 show only a slight decrease in population for the region's counties, but nearly as many families had moved back into the area after 1887 and prior to the census of 1890 as had migrated out during the drought. What was true during mid-century was still true: there were generally two streams of traffic in the southern plains, one entering during times of adequate moisture and the other exiting during drought. Throughout 1886 and 1887, the outward bound stream was significantly heavier.[63] The drought finally broke in May of 1887 with the arrival of typical spring rainstorms. Thus came and went the first major drought to visit the southern plains after its settlement by sedentary non-Indians, to be remembered by those settlers as a watershed event. Cattle prices rebounded from three dollars a head in 1886 to ten dollars a head after the drought had passed.[64]

Despite the dry conditions of 1886 and 1887, boosters continued to portray the plains as an area of promise. As in the 1830s, the boundaries of the Great American Desert were receding, in theory. As early as 1876, the Santa Fe Railroad promoters began stating that the "rainline" was moving farther west at the rate of eighteen miles per year, slightly preceding the advancing population.[65] Seventeen years and a major drought later, Union Pacific writers claimed in 1893 that "little by little, the Great American Desert [had] faded away."[66] J. B. Watkins, as a wise Kansas banker, refused to lend money for establishing a homestead west of the drought line, which he thought would not support farming, but even he was swept up in the optimism of his time. Watkins claimed that he was confident that "the rainless belt ha[d] retreated before the march of civilization."[67] Booster literature continued to attract settlers, even though the years immediately following the 1886–1887 drought were dry to average years in terms of moisture. The suffering and negative publicity of the intensely arid trend of the mid-1880s were not enough to curb promotional optimism or settlers' visions of opportunity in the southern plains.

By 1893 the southern plains would again be dominated by extremely dry weather, but this time the added burden of an economic crisis accompanied the meteorological phenomenon. The combination would prove disastrous for most farmers and ranchers in the region. The rising population on the southern plains meant that there were more people present to suffer the effects of drought, and all of them were sedentary.

The drought was preceded by a dry winter, and the high winds of spring brought a return of dust storms to the southern plains. The *Meade County Globe* reported that April 11, 1893, "was one of the worst days that we have ever seen in Kansas. The dust was thick in the air and drifted around in heaps like snow."[68] The arid conditions continued through the summer. A firsthand account of how the weather affected ranching in the far southern portion of the region is found in the diary of Robert S. Winslow, who ran a sheep operation just southeast of San Angelo. Ranching in this area relied heavily on windmills to pump water into tanks from which the livestock could drink. On July 11, 1893, Winslow claimed that there was plenty of wind to operate the water pumps, but there simply was not enough water underground to satisfy his stock. In fact, there were only two wells remaining that produced any water, and by late September, only the "Home Place" well provided any. Winslow had to sell much of his stock because he could not keep the animals alive, and the prices of their wool decreased because the dust storms dirtied the sheep's coats.[69]

Louis Hill, who worked as a cowhand in his youth and later ran a frontier supply store at Fort Griffith, Texas, moved to the growing Shackelford County town of Albany, Texas, in 1881. Hill formed a partnership with Sam Webb a couple of years later, and the two established a real estate firm to lease and sell land for their own profit and for speculators who lived within and outside state borders. The *Albany News* claimed that Webb and Hill did more real estate business than any other company in West Texas.[70] In 1893, however, the firm struggled. Prior to the drought, Webb and Hill had sold a lot of land, and the company collected payments from the mortgaged purchasers and sent them to the sellers. In August 1893 the two men found it hard to collect these payments. Earlier that year, a nationwide economic depression triggered by the Philadelphia and Reading Railroad's declaration of bankruptcy had created a tight money market. Investors in the rail companies began selling their stock to invest in the more stable commodity of gold, causing the gold reserve to fall and the stock market to crash. By the later months of the year, 600 banks had closed their doors, 15,000 businesses had failed, 74 railroad companies had gone belly up, and nearly 25 percent of the labor force was out of work. As if to make sure that no segment of the economy would thrive, an economic depression in Europe during the early 1890s drove crop prices down in the United States, increasing plains farm foreclosures dramatically.[71]

As a result of these conditions, many farmers found it hard to pay their debts. The drought had destroyed their crops and the depression had influenced local banks to curb lending, leaving southern plains farmers with little access to cash. Webb and Hill attempted to intercede on behalf of their neighbors by requesting extensions on the monthly payments. One such letter declared,

we have had the driest season since 1886—in fact it is *worse* than 1886, and it is preventing the leasing of many of your surveys. There is no grass and water—stock are suffering, and unless we have rains this month, all the stock will have to be moved out of this country, as later rains will not bring the grass in time for it to cure before frost. We are certainly having a hard time of it—the fearful depression in money matters, together with the exceedingly dry weather, is enough to drive the people half crazy.[72]

No rain had fallen in Albany between June 15 and August 4, when only a light shower teased the parched ground with moisture.[73] As the drought persisted into the fall, the struggle to locate water for area cattle became as acute in Shackelford County as it was for the sheep operators near San Angelo. Cattlemen began to unload their stock at the local sale barns and the price for beef again plummeted. Webb had made a trip to Graham, Texas, and on his way home found that he could not buy a bucket of water for his horse for the forty miles between the Brazos River and the town of Breckenridge. In Webb's own words, "the 1886 drouth was a picnic compared to the present."[74]

By September the firm found its own finances in question. Webb and Hill had also purchased thousands of cattle to run on leased or company land. Pasture conditions were so poor and the company's resources so strained that the firm advertised to sell all of their feeder cattle, some 1,500 to 2,500 heifers and steers, for $20 a head, and 1,000 cows at $9 each.[75] No buyers were found, and the partners began looking for other options, including leasing land in Indian Territory that still had adequate grass, arranging deals with feed lots to house their animals until it rained again, and even moving into other entrepreneurial opportunities, such as insurance sales.[76] By October Webb and Hill were in the unenviable position of asking their own creditors for extensions.[77] All through the winter of 1893–1894, Webb and Hill attempted to find a feed lot that could take their cattle, but the drought had left feed mills dormant due to lack of corn and the feed lots remained vacant because they had no water for the stock. Finally they were able to place one thousand head of cattle at a lot in Bryan, Texas, where rains had fallen in the spring. The drought had sapped the financial resources of Webb, who sold his share in the partnership to Judge J. A. Matthews in 1894. After solidifying his economic standing, Webb bought back his partnership later that year.[78]

Up in Oklahoma Territory, circumstances were similar. During the fall of 1893, Charles Alling set up a tree nursery in Perry. The treeless plains seemed the perfect location to sell seedlings, and indeed he had plenty of orders as new settlers arrived from the East and attempted to reconstruct their new environment to resemble their wooded homeland. Demand for Alling's seedlings outstripped his stock, requiring him to purchase $8,000 worth of additional plants. Although he sold $12,000 worth of seedlings, Alling's venture met with failure, for he failed to consider the weather. The ground was so hard by 1894 that settlers could not dig holes for the plants. Furthermore, the lack of rainfall would force farmers to use precious water on trees when it was needed for their own and their livestock's consumption. Three carloads of seedlings were returned to Alling because the shipping fees were not paid in Lawton and Anadarko, Oklahoma Territory. In a last-ditch effort to save his enterprise, Alling borrowed the city of Perry's water pump with the intention of irrigating his trees, but there simply was not adequate subsurface water available for the roots to tap.[79]

As in the previous drought of 1886, there were efforts to obtain and provide relief. In 1893 Greer County, Oklahoma Territory, accumulated only 11.67 inches of rain in an area that averages 24 inches a year.[80] Homesteaders quickly used up their reserves of food. Without crops, grass or water, they were not able to maintain their small cattle herds or produce vegetables for family consumption. These people were faced with starvation as the drought lengthened. As in the previous drought, a minister took on the mantle of drought relief spokesman. A local preacher named Mr. Kizzar volunteered to travel to Fort Worth to solicit aid. Kizzar spoke at several churches in Fort Worth and described the serious plight of the pioneers

in Greer County. The people of Fort Worth responded with enough donations to purchase food and ship it by rail to Quanah, Texas, thence by
wagon across the Red River.[81] Many people also took advantage of the government's policy to issue bacon and beans to those in need from military
posts.[82]

Statistics for 1894 reveal the severity of the continuing drought. The
U.S. Department of Agriculture established a weather bureau that year that
tabulated climatic data at four sites throughout Oklahoma Territory. The
records for Guthrie, Logan County, show that 22.99 inches of rain fell in
1894, representing only two-thirds of its average precipitation. Further east,
the *Lincoln County Recorder* mentions only 22.39 inches in Meeker, which
usually receives 36 inches. Mangum reported 8 more inches than the previous year, but this was still significantly below normal. The town of Jefferson, Grant County, recorded the most severe shortage of rainfall that year.
Only 18.29 inches of rain fell, a little more than half of the 32 inches the
area typically receives.[83]

By the third year of this drought, the whole of the southern plains
was ripe for wind erosion. Eighteen ninety-five was truly the year of dust
storms. The two preceding years of drought, combined with the desperate farmers' practice of grazing their herds until they could find a buyer
for their cattle, had again denuded much of the region of ground cover.
When the winds picked up, as they invariably do, they would lift millions
of tiny particles of soil and carry them hundreds, sometimes thousands,
of miles before the gusts slowed, depositing the sand and dirt back on the
earth. J. C. Neal of the Oklahoma Agricultural and Mechanical College in
Stillwater, reported on January 20 that at 9:00 A.M., the winds began gust-
ing up to fifty-five miles per hour and the temperature had fallen rapidly.
Neal reported that as flashes of lightning lit the sky, he was awestruck to
see a cloud of dust reaching a phenomenal one thousand feet into the atmosphere.[84] Another severe dust storm hit Enid, Oklahoma Territory, on
March 19, with winds of up to eighty miles per hour, completely suspend-
ing travel in the vicinity as the air was filled with blowing dirt. After the
storm, the wheat and garden crops of local residents were "hidden from
view under several inches of dust."[85]

As if to prove wrong those residents who had optimistically hoped the
worst was over, the month of April was even more beset by blowing sand.
El Reno, Oklahoma Territory, reported a storm on the 6th. Alva, Healdton,
Ponca City, and Pond Creek all communicated dust storms on the 14th and

15th. The most notorious storm hit western Kansas on April 12, 1894, when an intense cold front was preceded by strong winds that blew loose dirt before it, forcing those in its path to endure a ferocious dust storm that piled up to six inches of sand along nearby railroad tracks. This was followed by a deadly blizzard. Five thousand cattle perished as they put their backs to the storm and made their way downwind. The dirt and snow piled up in the draws and creek beds, obscuring the dangerous pitfalls awaiting any animal that tried to cross; after the snow had melted, countless carcasses lay in these ravines.

The most poignant tale associated with this weather phenomenon occurred near Johnson City in Stanton County, Kansas. As the storm approached, Ma Dick sent her ten-year-old son, Charlie, to round up the family's livestock to protect them from the impending weather, which appeared to be only another dirt storm. As he was preparing to embark on his chore, his little eight-year-old sister, Cora, begged Charlie to take her with him. The seemingly benign decision to allow the girl to go with her brother turned out to be fatal. Their mother reported that shortly after they left, the storm hit and raged for forty long hours. The children's father had been away in town doing business and was understandably delayed in returning to the farm. On his way home the day after the blizzard, he saw the family horse standing beside the road with two huddled figures at its feet. When he neared the bodies, he identified them as his own children, lying with their faces together and their arms wrapped around each other in a loving embrace, just as he had seen them many times sleeping together in the warmth of their home.[86]

There was another fatality during that storm. Bertie Orth, a thirteen-year-old cripple, also rode out to bring in his family's cattle with his crutches tied to the saddle. It was the last time he was seen alive. Bertie's body was found just half a mile from his home. He either fell or was thrown from his horse and had attempted to crawl back to shelter, but eventually he succumbed to fatigue and overexposure, his jeans worn through, exposing knees rubbed raw from his efforts.[87] The environment had dealt a painful lesson to those who wished to homestead on the plains. Drought and blizzard tested the resolve of the pioneers, breaking the will of most to remain in a region of such tempestuous extremes.

The length of this drought caused a widespread reaction. After witnessing farmers struggling to harvest crops for years, the editor of the *Kingfisher Free Press* in Oklahoma Territory claimed that wheat farming was "a fake and a fraud."[88] Just west of Kingfisher, the *Watonga Republican* had a completely different message,

> Don't sacrifice your claim now, this is an unusual experience Oklahoma is passing through. Perhaps never will come again. Most of you were not able to own a claim elsewhere. You came here to get one. Where can you get another if you sacrifice this? The outlook may be dark, but it is just as bright as it was three years ago when we set our stakes in Blaine County. . . . Three Kansas and Nebraska farmers have just recently taken advantage of the present scare and bought at much less than they would have had to pay just six months ago, and Oklahoma is just as good as it was then. Brighter days will come again.[89]

Such optimistic words could not overcome the reality of coping with the vagaries of the weather. Once again, thousands of wagons were heading east.

Richard S. Cutter of Ochiltree County in the Texas Panhandle was one of those who took his family east. The Cutters had migrated from Ness County, Kansas, to Texas on March 14, 1887. During the six years prior to the drought of 1893–1896, Cutter had developed a crop rotation that worked marvelously. In May he and his sons planted cane; from mid-May to June they planted millet; then, in late June they harvested wheat or rye. During July they harrowed the wheat and rye fields. By August they harvested millet, and in September, the cane. Then the Cutters prepared the land for sowing wheat and rye and tended to their cattle through the winter. Cutter surely thought he had found his own little piece of heaven out on the high plains of Texas. Then the weather changed. The family continued this crop rotation during the dry year of 1893, but their yields were poor. By March 1894 their income was suffering, but they still managed to lend ten dollars to friends in Oklahoma Territory, who seemed to be worse off than they were.[90]

The summer of 1894 was extremely hot. Cutter's crops withered in the scorching wind, but when the wind stopped blowing, the windmill could no longer pump water for the livestock. Cutter decided to forego planting another crop in the baked soil to take a trip to visit friends. Thus on July 13 when the family usually prepared to harvest millet, they simply packed up their wagon and traveled to Harvey County, Kansas, for a two-and-a-half month tour of their friends' homes in McTearson, Rice, and Ness Counties.[91]

In 1895 the prospects for farming were no better. The land remained too dry to plow until May 29, when a shower moistened the soil, but by July 17 the hot, dry weather had practically destroyed the millet crop yet again. This time the Cutters joined the exodus and traveled from the Texas Panhandle to Effingham County, Illinois, where they stayed with relatives from September 24 to April 29, 1896. Many families who owned land on the southern plains during this particularly dry trend reacted in a similar manner. Others totally gave up and abandoned their claims altogether, their dream of farming broken on the plains. It is estimated that 50 percent, or more, of southern plains residents vacated the region afflicted by this drought. In a few counties, the depopulation was complete. In terms of proportion of population, the mass movement of the late nineteenth century dwarfs that of the famous Okie migration along Highway 66 during the Dust Bowl years, but the former exodus was generally to the east rather than west.[92]

The ever-present triumvirate of fires, high temperatures, and dust storms accompanied the drought of 1896. On March 13 a "disastrous prairie fire" struck near Hardesty in Beaver County, Oklahoma Territory, consuming the stock, grasses, and buildings of five area farmers. One brave woman who was at home alone, Mrs. Carter, fought the blaze for hours and was fortunately able to preserve her house with the furniture and family heirlooms as well as the livestock. The *Hennessey Clipper* reported that "she was found lying on the prairie unconscious and painfully, but not fatally, burned."[93] The temperatures were also unusually hot for late spring. The *Hardesty Herald* reported on June 8 the highest temperature to hit the area in six years, 103 degrees Fahrenheit. The newspaper also stated that local farmers had already turned the wheat crop under because it showed no promise for harvesting, and they had replanted the fields in corn. The hopes for this second crop, however, were slim. Furthermore, there was "no water in plots or holes, or on Peon Creek—unprecedented in the past six years."[94] The week of September 6, an intense dust storm hit El Reno, Oklahoma Territory. According to the local newspaper, "it [was] almost impossible to see the business houses on either side [of the streets] for the dust." The concerned author advised purchasing sprinklers to wet the town streets in order to reduce the effects of future dust storms.[95]

Non-Indians were not the only people affected by the drought; the Natives in Oklahoma Territory were perhaps even more susceptible to the vagaries of the weather. Non-Indian settlers could pack up and move east to locate jobs or move in with relatives until the dry spell had passed, but tribal members were either trapped on a reservation or living on an allotment with no opportunity to move away from the drought-stricken area.

Along with the handicap of a fixed location, the Native Americans' way of life was under attack. In 1884 a group of reform-minded people met at Lake Mohonk, New York, to discuss the future of the Native people. These men and women considered themselves friends of the Indians, who, in the minds of the reformers, needed to be saved from themselves. The best way to protect Natives from their traditional communal culture was to destroy the reservations, where the tribes held land in common. In order to erase the communal spirit of the Indians and speed them along the way of "civilization," these reformers called for the tribal domains to be divided into individual allotments. This proposal to dissolve the reservations also appealed to those who were certainly not friends of the Indians: western farmers, real estate speculators, and railroad companies liked the idea of allotments as well, for this strategy would leave millions of acres of land unclaimed by anyone. They hoped these lands would become part of the public domain.

In 1887 Senator Henry Dawes of Massachusetts won the passage of the General Allotment Act. According to the legislation, reservation members could claim an allotment of one hundred sixty acres if they were a head of a household. Orphans and single males could qualify for allotments of eighty acres. To accept an allotment, all a tribal member had to do was sign the tribal roll, and upon receiving his allotment, he would become a citizen of the United States—but often without voting rights, which were restricted by the states. Those who refused to sign the roll sheets either languished on the diminished reserves or had allotments selected for them. The process of securing a tribal roll, surveying the land, and assigning allotments occurred earlier among some tribes than others. During the drought of 1893–1896, the Cheyenne and Arapaho, Wichita, Caddo, Iowa, Sac and Fox, and Kickapoo tribes had already been allotted lands, while the Tonkawa, Ponca, Otoe, Comanche, Kiowa, and Kiowa Apache tribes had not yet divided their domain into single holdings.

Almost immediately after the Cheyenne and Arapaho tribes went through the allotment process, the drought hit the reservations. In September 1894 the agent at Darlington, A. E. Woodson, reported to his superiors that the dry weather had "a discouraging effect on the Indians. The scarcity of water (the streams all being dry) ha[d] served to bring them together in large camps at points where they could obtain a supply sufficient for all purposes."[96] The drought was obviously impeding attempts to "civilize" the Natives. How could the allotment system work if the Indians continued to gather at central locations to find water? The drought affected other tribes as well. The Osages' corn crop was a failure in 1894, as was that of the Tonkawa.[97] The Poncas likewise reported a yield of only six bushels per acre of corn.[98]

As the drought continued into 1895, the superintendent at the Cheyenne Boarding School reported complete failures of its wheat, oats, potatoes, early corn, and garden vegetables. The region must have experienced some early summer showers, for the students were optimistic about their late crops of cowpeas, Kaffir corn, and prairie hay, but these would not be harvested until September.[99] The absence of water caused great concern for W. J. A. Montgomery, the superintendent of the Arapaho Boarding School at Darlington, Oklahoma Territory. The creative director ordered the construction of two cisterns with a combined storage capacity of eight hundred barrels of water. Montgomery claimed that this alteration of their waterworks system was possibly "the most important improvement made during the year."[100] Students at the Kiowa Rainy Mountain School, forty miles west of Anadarko, hauled water from a spring three miles away until they dug a new well closer to their dorm.[101] On the Ponca Reserve, a spring dust storm ruined the corn, oats, and garden vegetables, blowing off the entire topsoil "to the depth of plowing."[102]

During the fourth consecutive year of drought, the Indian agents' frustration with trying to maintain their agencies and schools, while at once demonstrating their wards' improvements toward "civilization," mounted. The Cheyenne agency invested in a new windmill and set it up with high hopes; but in the words of A. H. Viets, "the pump is an excellent one, and if the water were there, it would handle enough to supply three plants such as this one."[103] The agent went on to state that he had been vexed constantly with maintaining a water supply for the Cheyenne School, where the cistern at full capacity contained only three-fourths the water necessary to run the school. Of course, bathing was a low priority, and Viets allowed the students a bath every fourteen days, alternating the privilege to the girls one week and the boys the next. During the warmer months, he sent the male students down to the river, four miles away, for their ablution, thus saving the school's water, but according to Viets, "when there is nothing but dry white sand in the river bed, the bath, although it may be hygienic, is not especially refreshing."[104]

The Indians were depressed by the difficulty of farming during a drought as well, particularly the Plains tribes, who were no longer allowed to enjoy their nomadic lifestyle. For more than a century and a half, they had ranged over the southern plains, specializing in hunting buffalo and trading for vegetable products with the Pueblos or the Caddos. After their removal to reservations, their agents urged them to take up agriculture, and many did, even though it was considered women's work in their culture. Allotment entrenched the necessity of participating in land cultivation as government annuities became a thing of the past. But it was not until 1896 that the Oklahoma Agriculture Experiment Station began studying the possibilities of irrigation in Oklahoma Territory, and the agricultural lifestyle remained difficult.[105] The extended drought greatly demoralized members of the Cheyenne and Arapaho tribes. Environmental conditions assured their failure at every turn, yet they were stuck on their one hundred sixty–acre allotments. It is a cruel irony that the sedentary non-Indian settlers could move to find work or relatives who would help them out financially, while the once-nomadic Natives had to remain and simply endure the hardships. The Cheyenne and Arapaho agent, A. E. Woodson, claimed, "A laudable disposition has been shown by a very large majority of the Indians to cultivate their allotments, but the prevailing drought of the past two seasons has had a very discouraging effect on their efforts in this direction. The wheat and oat crops were a total failure."[106]

The drought had led both Indian and non-Indian families to curtail traditional agriculture on the southern plains and turn to ranching. The Cheyenne and Kiowa agents suggested that "extensive farming . . . be given up."[107] Thirty-four consecutive days with temperatures averaging 107 degrees Fahrenheit had destroyed any hopes of raising a crop, and this was the fourth year of these circumstances. Many non-Indian farmers, experiencing the same climatic conditions, reduced their crop land and increased their pasture areas to accommodate more cattle. They had noticed that the native grasses were much hardier and survived the extremely dry conditions that had destroyed their crops.[108]

Between 1890 and 1900 there was a notable population decrease in the southern plains counties of Kansas and Colorado. Of thirty-two Kansas and Colorado counties affected by the drought, twenty-four decreased in population during these years. The counties that maintained their populations did have something in common: seven of the eight counties bordered the Arkansas River and also lay along the route of the major rail line of the region. All of the southern plains counties had experienced incredible growth during the previous decade, rising in population from a combined 52,958 to 122,768. This number then fell to 100,465 by 1900.[109] In addition to the drought of 1893–1896, the 1893 Cherokee Outlet Land Run in northern Oklahoma Territory certainly drew its share of settlers away from Kansas, as all the counties that bordered Oklahoma Territory saw population losses. Emigration had accompanied the two previous droughts of the century, as people headed east or west out of the Plains to find subsistence. Just as the Cutter family loaded up their wagon and headed east to visit friends and family during the months of July, August, and September, when they usually harvested millet and cut cane, countless other families crossed the prairies in search of jobs or friendly faces.

Participants in the Cherokee Outlet Land Run and their experience on the plains over the following four years provide the perfect example of the drought's effect on farm settlement. The booster voice had, like a siren's song, lured many farmers to the southern plains, but there they encountered the last major dry trend of the nineteenth century. Without the benefit of organized governmental aid, farmers had to fend for themselves or seek relief from neighbors who were often experiencing the same conditions. With the choice of either moving or starving, most decided to leave the region at least until the drought subsided, if not forever. The boom-bust cycle of the frontier had once again left its mark on the plains.

TWELVE

Attempts at Rainmaking

Any party which takes credit for the rain must not be
surprised if its opponents blame it for the drought.

DWIGHT MORROW

L. Frank Baum had hit on a run of bad luck, beginning with the mis-
adventure of attempting to keep his father's business afloat in New
York. When an embezzling accountant left the company broke, Baum
headed west for a new start and opened a general store in Aberdeen, Dakota
Territory. Misfortune followed him there: the severe drought of 1886–1888
withered the crops and settlers' hopes. His dream of operating a general
store evaporated with the massive outmigration from the plains. Not one
to be deterred easily, Baum edited a newspaper in Aberdeen as the town
tried to persevere through the dry trend, but the nationwide economic depression reduced the number of subscribers and advertisers. Once again,
Baum packed up and moved, hoping to put the past behind him and begin
anew, this time in Chicago. In 1893 the Columbian Exposition blew into
the Windy City and promised a renewed optimism in the inventive genius
of U.S. citizens. Although he landed a job as a traveling salesman, the one
thing Frank Baum excelled at was writing, and, moreover, he had a gifted
imagination for telling children's stories. Baum put his talent to work during his off time and produced his most famous work, *The Wizard of Oz,*
which he published in 1900. The hopes of Baum and thousands of others of starting anew in the West had met with disaster due largely to the

late nineteenth-century droughts. As a resident of the northern plains, he experienced many of the same circumstances that befell southern plains denizens during those years.

While most residents of the plains found it difficult to subsist on the parched grasslands, area farmers were especially hard hit by the successive dry years. Always a risky venture, agriculture west of the ninety-eighth meridian was even more a gamble, and the odds in favor of the farmer decreased as he moved west. At the close of the Civil War, the relationship between landowner and worker had changed forever in the southern United States. The old plantation system was no longer possible after the Thirteenth Amendment abolished slavery. There were still plenty of laborers, but they would not work as chattel. In the new system that emerged, landowners allowed farm families to use sections of their land. All such families needed to bring to the agreement was their muscle and willingness to work; the landowner provided the mules, implements, and seed. In return for these benefits, the farm family gave the landowner a share of the crop raised in the field, earning the system the moniker "sharecropping."

As landowners repeatedly provided seed only for cotton, the single crop that had a chance of bringing a good return on their investment, the quality of the land deteriorated, and the price of cotton plummeted. For the first ten years after the war, the price for cotton fluctuated between a viable twelve cents to eighteen cents per pound, but after 1875, increased production pressed the price down to eleven cents a pound. Cotton farmers would not see a higher price for the duration of the century. On most farms, a pound of seed cost eight cents, not considering the other expenses in producing the crop like fodder for the mules, implement repair, and freighting costs to haul the crop to the gin. By 1885 cotton had further fallen to nine cents per pound, stressing the nerves of both the landowner and the workers.[1]

Small independent farmers suffered under these prices as well. Finding themselves strapped for cash, many farmers had to charge purchases to their accounts, allowing store owners to use the farmer's potential crop as collateral for a debt that, chances were, could not be paid in cash. Thus in this crop-lien system, the farmer's crop was pledged to the store owner before it was ever harvested. Making matters worse, store owners often inflated prices for those indebted to them, taking ever greater portions of the crop for themselves and leaving the farmer with less money to start the next year. As the price for cotton remained deflated, many small farmers could not hope to repay their debt to store owners and lost their land as a result. Now the best they could hope for was to become tenant farmers on someone else's land, giving up a share of their crop, or to move out west and start over. As we have seen, railroad promotions portrayed the plains as an extremely attractive location to do just that.

Yet the railroads presented another set of frustrations. Commercial agriculture was arduous without the rail services. Husbandmen would have to haul their crops to the nearest station or to a mill for processing before they could sell it, feeding their team of horses or mules along the way and losing time at the farm, which cut deeply into profits. Once the rails extended to a farmer's vicinity, selling surplus produce became an option, but often these rail companies had a monopoly on their services in the region. Their fees could be as high as the cost of freighting one's own crop by wagon. Such was the exacting dominance of the railroads on farmers that in 1901, journalist Frank Norris wrote an eponymous novel comparing the rail lines to the tentacles of an octopus, squeezing the life out of struggling farmers.[2]

Another perceived threat to agriculturalists was known as the "Crime of '73," when the Treasury Department took the silver backing out of the nation's printed currency. During the Civil War, the federal government had created so-called greenback bills to represent so many ounces of gold and silver that backed them from the federal reserves. Banks supported a bullion-based coin currency, since many wealthy northeastern businessmen had invested in war bonds during the Civil War when silver and gold backed the currency, but with the gold standard, they could redeem their bonds at a significant profit. For small farmers, the gold standard made it more difficult to come by hard cash, as the bills in circulation were withdrawn and replaced with far fewer greenbacks or gold coins.[3] Paying off debt to store owners became nigh impossible. In response to the hopelessness of the crop-lien system, the stranglehold of the railroads, and the burden of the gold standard, a group of farmers met at the farm of J. R. Allen in Lampasas County, Texas, in September 1877. Calling themselves the Knights of Reliance, members hoped to educate themselves on how to overcome these seemingly insurmountable obstacles to maintaining a family farm. By the next year, they had changed their name to the Grand State Farmers' Alliance, which people soon shortened to the Farmers' Alliance.[4]

Spurred on by the colorful speaking of S. O. Dawes and the organizational acumen of William Lamb, the concept of forming local Farmers' Alliances spread, first through Texas, then to the rest of the South and eventually north into Kansas. These alliances became popular throughout the South and the West, since farmers were so frustrated with the economic system they were bogged down in and political leaders so much more susceptible to the influence of economic interests than their constituents' welfare. Farmers gathered in the late summer to hear Mary Lease, a Populist activist from Kansas, speak on organization. It seemed to her that if farmers had half a brain they would organize and begin calling for reforms. To the small farmer, many things seemed unfair. The inordinate amount of influence eastern businessmen had on federal decisions, the onerous crop-lien system that forced many off of their ancestor's lands, the gold standard, the seemingly exorbitant railroad fees, and droughts: all conspired to take away any hope of maintaining a family farm.

Let us not forget the drought, which appeared to fly in the face of the pervasive later nineteenth-century view promoted by real estate speculators, railroads, and town boosters that the Great Plains was a garden. If only there were a way to ensure rainfall, the region would be a farmer's paradise. Certainly, people of all cultures and time periods have tried to find ways to induce rainfall. A *National Geographic* article published in 1894 stated that in Massachusetts, folklore claimed that simply killing a frog could bring rain. In South Carolina, the folktale seems almost as humorous, as the task of creating rainfall there is herculean. According to the old-timers, all one had to do was "catch an alligator, tie him to a tree, and whip him to death."[5] A series of more modern, "scientific" theories sprouted from the need for rain. Although none proved completely successful, they reveal how some residents of the plains attempted to cope with their arid environment.

In 1873 a young University of Nebraska biology professor named Samuel Aughey, who had only been teaching at the institution for two years, stepped into the limelight of the rainmaking debate by providing a scientific explanation for how the plains would be conquered. On January 20 he addressed the Nebraska state legislature, endorsing a theory known as "Rain Follows the Plow" and arguing that "one of the most interesting of the meteorological facts which affect [the plains] is this—that as civilization extends westward the fall of rain increases from year to year."[6] By 1880 the theory was widely embraced, not surprisingly, in the western United States. Aughey might have explained this theory like this: Cultivation of the soil allows it to retain more moisture by making it more porous, and the increased vegetative cover associated with agriculture releases more moisture into the atmosphere through transpiration than ever before. Clouds then form over new croplands, resulting in increased rainfall. With advocates like Aughey, the theory seemingly had scientific backing—always an effective promotional tool—and, of course, people in the plains wanted to believe rainfall could be encouraged on the plains. The theory brought hope to farmers that the future would bring ample rainfall; to real estate speculators, who saw the theory as a means of convincing would-be settlers that Americans were overcoming the Great American Desert with each crop planted; and to railroad companies, which, crossing their fingers that settlement of the plains would increase demands for their services, contributed substantial resources to promote the theory as well.

The following poem, published in the *Watonga Republican* of Indian Territory on May 5, 1895, aptly describes the theory:

I heard an old farmer talk one day
Telling his listeners how
In the wide new country far away
The rain follows the plow

As fast as they break it up, you see
The heart is turned to the sun
As the furrows are opened, deep and free
The tillage is begun

The earth grows mellow and more and more
It holds and sends to the sky
A moisture it never had before
When its face was hard and dry

And so, whenever the plow shears run
The clouds run overhead
And the soil that is stirred and lets in the sun
With water is always fed

I wonder if that old farmer knew
The half of his simple words
Or guess the message that eternally true
Hidden within it was heard

It fell on my ears by chance that day
But the gladness lingers now
To think that it is always God's own way
That the rainfall follows the plow.[7]

The year this poem was published, ironically, was a drought year. Support of the theory continued through the severe droughts of the late 1880s and the mid-1890s, long after thoughtful people should have realized that rainfall was not increasing.[8] It was faith in the concept of "Rain Follows the Plow," however, that convinced many non-Indian agriculturalists to move out onto the high plains to try to farm it. Despite the many people proclaiming milk and honey on the plains, the fact is that the period from 1870 to 1899 was average to dry.

Another theory on human-induced rainfall popular during the 1890s can be traced astonishingly all the way back to Plutarch's belief that rains followed major battles. In his *Life of Marius,* Plutarch claims: "And it is said that extraordinary rains generally dash down after great battles, whether it is that some divine power drenches and hallows the ground with purifying waters from Heaven, or that the blood and putrefying matter send up a moist and heavy vapour which condenses in the air."[9] Many in the late nineteenth century noticed that precipitation seemed to follow Fourth of July celebrations and Civil War battles. This observation led to several experiments at producing rainfall by detonating gunpowder in the sky. Concussion theory, as it was known, held that the combustion of explosives in the heavens shook the atmosphere in an "air-quake" that compressed water molecules, forming rain droplets heavy enough to fall to the earth.

In 1871 Edward Powers, a civil engineer from Chicago, published *War and the Weather; or, the Artificial Production of Rain,* describing the relationship between military battles and rainfall.[10] Then, in 1880 Daniel Ruggles patented a method for obtaining rainfall through the use of dynamite and added the novel idea of releasing explosives from hot air balloons to more effectively carry out the project.[11] Ruggles lobbied U.S. Senator Charles Farwell of New York to present a proposal to fund the testing of this theory. In 1874 the senator brought the proposal up, but Congress failed to take any action on it. Senator Farwell continued working for the project, pledging his own money to support the tests. After the usual lengthy debates, Congress appropriated $2,000 to test the theory in 1890, then increased this sum to $7,000 later that year. In all, Congress reserved about $14,000 to support the research of Major Robert Saint George Dyrenforth, a Civil War veteran and lawyer from Chicago who worked in the U.S. Patent Office.[12]

In the summer of 1891, Major Dyrenforth's first experiment proposed testing the effects of explosives at three levels of altitude. A wealthy Chicago meatpacker named Nelson Morris offered to host the experiment at his C (for Chicago) Ranch north of Midland, Texas, which had suffered eighteen months with no rainfall. Dyrenforth planned a three-pronged attack on the West Texas atmosphere, utilizing conventional military artillery to concuss the lower levels. To test the mid-levels, he constructed sixty-eight cloth kites, harnessed them with explosives, and connected them with an electric wire coupled to a detonator. Finally, to reach higher altitudes, Dyrenforth ordered the release of sixty-eight explosive balloons, ranging between ten and twelve feet in diameter, from three larger hot air balloons. Folks at Midland must have thought the carnival had come to town, but the results were less than spectacular. A slight shower followed the test, though according to one observer, it was not worth the money and effort put into the project. Regardless of the naysayers, Dyrenforth claimed a victory and his dispatches to Congress painted a promising future for concussion theory. However, in his report, Dyrenforth hedgingly claimed that when conditions are favorable for rain, the application of concussion theory could generate rainfall, but when circumstances are not so optimistic for showers, concussion constitutes a "wasteful expenditure of both time and material in overcoming unfavorable conditions."[13]

Town leaders of El Paso heard of Dyrenforth's success and were desperate enough to pay his team to visit their dehydrated city. Once again, the findings were inconclusive. An aging Robert Kleberg offered a hefty sum for the researchers to test their theory at the King Ranch. Dyrenforth set up Camp Edward Powers, named after the man who wrote the book on the concussion theory, and after his tests, rains showered the dusty ranch. A now thoroughly convinced Congress added another $17,000 worth of funding to the experiments.

The last of the Dyrenforth experiments occurred at San Antonio, beginning at 4:00 P.M. on Thanksgiving in 1892. The sky was overcast, inviting Dyrenforth's efforts to draw forth droplets from the clouds. The crew released balloons into the foggy sky, losing sight of them until they exploded within the clouds, creating quite a display for onlookers. Professor A. Macfarlane from the University of Texas, a professed skeptic, attended the experiment as an uninvited guest and had little good to say of the whole affair. After more than seven hours of explosives concussing the atmosphere, the professor claimed the "mist ceased, and the stars appeared in places nearly overhead."[14] Following this well-publicized failure, the disappointed public lampooned the major, jokingly referring to him as Major "Dryhenceforth."[15]

The theory was not completely lifeless but re-emerged twenty years later. Charles "C.W." Post, famed for dry cereals like Elijah's Manna, later renamed Post Toasties, purchased two hundred thousand acres from the Double U Ranch and set about establishing a model community forty-five miles southeast of Lubbock, to be named after its founder. Like other boosters before him, he had to overcome the dry tendencies of the area's climate to lure settlers. Thus he was keen to retest the concussion theory. The Texas and Pacific Railway also conducted a concussion test near the coal mining town of Thurber, along their line. Both experiments proved inconclusive.[16]

Any discussion of rainmaking would be incomplete, and less colorful, without discussing the rainmakers who claimed they had a chemical formula that could induce the moisture to fall from the sky. Many towns and farmers fell victim to these charlatans, and no wonder, since farmers had poured their sweat into breaking and planting the fields, only to stand hopelessly by and watch their crops slowly burn to yellow. Prayer sessions at local churches looked to God for rain. When their petitions proved unavailing, some people became desperate enough to turn to any scheme that promised precipitation. Enter the Rain Wizard.

One of the most notable rain wizards was an Australian named Frank Melbourne, who claimed to have a machine that could produce moisture. It appears that Americans' susceptibility to a clever sales pitch, especially when accompanied by an Australian accent, is a phenomenon that knows no boundary in time. Although people loved to listen to Melbourne talk, he would absolutely refuse to discuss his machinery and chemical formula for producing precipitation. But, for a fee, he would travel to any town and work his magic. Folks heard tales that he could even dictate the day and hour of his showers, avoiding raining out a baseball game in Ohio, then opening up the heavens for a real toad-drencher. After a successful rainmaking in Cheyenne, Wyoming, the townspeople of Goodland, Kansas, suffering from dried fields and withered wheat, hired Melbourne at a hefty price of $500 to produce rain. The city fathers felt it was a safe bet, for if the wizard failed to bring precipitation, they would not be required to pay. If, however, he did cause the heavens to open, they would reap harvests worth more than their initial investment.[17]

DATES,

SEPTEMBER

22nd to 26th

Inclusive

Frank Mel'ourne, the Rain Wizard.

THE MOST
Wonderful
Inventor
of the
Century.

Melbourne the "Ohio Rain Wizard"
Will be at GOODLAND

Fair Week.

And has contracted to produce a
Heavy Rain the last day of the Fair,
Sep't 26th.

Governor Humphrey

Will be present one day.

The management of the Fair Association will spare no pains or expense to
make this fair the most entertaining
of any ever held in western Kansas.

One and One Third Rate

has been secured over the Rock Island.

Alex Martin, Pres.,   Wm. Walker Jr. Sec'y.

SHERMAN COUNTY FARMER Print, Goodland, Kansas.

"Melbourne the Ohio Rain Wizard" handbill, *Sherman County Farmer,*
September 17, 1891. Courtesy of kansasmemory.org, Kansas State Historical
Society.

Melbourne arrived on September 25, 1891, as the summer drought had continued into early fall. Town merchants sprang for the construction of a two-story building near the fairgrounds for the rainmaker to set up his laboratory. On the second floor the rain machine was placed precisely below a carefully crafted hole in the ceiling, so that the professor's gasses could rise into the atmosphere and create the chemical change needed to produce rain. The first floor housed Melbourne and his brother, whose main duty was to keep people out of the lab. At Goodland, the famed Rain Wizard's efforts did not meet with the same results as his reported displays of meteorological wizardry in Ohio. Although his method had failed to bring rainfall to the community, he still left town with a jingle in his pocket.[18]

Local entrepreneurs in Goodland were so impressed with Melbourne's method that they convinced him to sell his concept. With their new rain-making patent, the investors formed the Inter-State Artificial Rain Company (ISARC), an impressive name to match the hoped-for profits. The company quickly sent agents to conduct tests in Oklahoma Territory and in Texas. On October 27, 1891, a telegram from Oklahoma City claimed the ISARC method had produced the first rain to hit the area in six weeks. On November 1, a similar report came from Temple, Texas. ISARC agents claimed the tests were so successful that members of the firm sold their concept to a joint-stock company for fifty thousand dollars.[19]

The ISARC's success drew imitation. By 1892 two more rain companies sprouted out of the dry plains soil of Goodland, Kansas: the Swisher Rain Company and the Goodland Artificial Rain Company. Individual entre-preneurs also saw opportunity in researching means to create rain. Clayton B. Jewell, a Goodland native and chief dispatcher for the Rock Island Rail-road Company, latched onto the notion that rain followed telegraph poles. In 1876 French scientists in Algeria had claimed to create rainfall by using kites to create an electrical connection to clouds at four thousand feet in the atmosphere. The resulting fog promised a future in harnessing electric-ity as a conduit for rainfall.[20] Professor Jewell, as he liked to be called, was an expert electrician and his experiments used electric batteries as well as balloons and explosives to test concussion theory. After an inspiring visit with Frank Melbourne, Jewell switched theories on pluviculture and began to invest most of his funding in procuring chemicals.[21] The Rock Island provided the professor with all the needed supplies and a special railcar to house his lab, replete with bottles, tanks, and a pipe to conduct the steaming chemicals out through the roof of the vehicle. The company further allowed him to travel along their lines free of charge. In fact, Jewell was a rainmaking agent for the Rock Island, performing his tests free of charge to communities along the rail company's lines.

This pro bono work, of course, dug into the other artificial rain companies' profits, but it also kept Jewell's name in the local newspapers. On June 1 and 2, 1893, the professor was unable to induce sufficient rain at Meade Center, Kansas, to claim a successful test, but he did predict that the high winds that ruined his efforts would blow his chemicals in a northeasterly direction and cause a sizable rainshower around Salina, Kansas. Heavy rains hit Salina the next day, and although his planned test had been a failure, Jewell was legitimized in the view of many people. On June 6 the professor was in Dodge City, Kansas, mixing chemicals, but he was unable to procure a shower due, he claimed, to high winds.[22]

By the next month, Jewell had moved on to Oklahoma Territory, where he was providing free tests of his artificial rain–making method. The *Beaver Advocate* gave him some free publicity in its July 27 issue, in which it claimed, "Rainmaker Jewell brought down the stuff at Duncan, O. T., Sunday. He sprinkled down a territory seventy-five by one-hundred miles in extent."[23] Jewell traveled to Hennessey, Oklahoma Territory, for the opening of the Cherokee Outlet Land Run, but was unsuccessful at inducing rainfall. However, he did purchase a hot air balloon and sold rides for one hundred dollars a pop to those wishing to claim the choicest plots. For this sizable fee, a run participant could forego the dusty, mad dash into the Cherokee Strip and from the air, pick out the best land near a water source. Jewell would then lower the hot air balloon and allow his client to be the first to claim the land.

Early in 1894, the Rock Island increased its funding for Jewell's experiments, but by July interest in rainmaking had waned. The drought had defied the rain wizards and had broken the public's faith in their abilities.[24] The last anyone heard from the professor, he was in Los Angeles, California, at the behest of some desperate farmers, but after he was only able to raise a few clouds, Jewell refunded his fee and returned to Kansas.[25] Frank Melbourne fared even worse. He continued to sell his rainmaking services throughout the West, but some eavesdroppers found that he had consulted the *Farmer's Almanac* for likely dates of precipitation to determine where his best chances of rainmaking success might be. He took his own life in a Denver motel.[26]

A good illustration of the drought's emotional and financial devastation on many families can be observed in the life of pioneer Max Krueger, a German immigrant who came to Texas in 1868 at the age of twenty-one. When Krueger was a child in Germany, his father died, leaving his mother to try to provide for her children. She ensured that Max received an education up to the age of fourteen, at which point he became an apprentice in a silk manufacturing house in Berlin. The cold Berlin winters were hard on the young man's health, and physicians advised him to seek a warmer climate. At the end of his apprenticeship, Krueger accepted a position as a traveling companion and interpreter for a family journeying to Le Havre, France. Krueger had taken some French in school and had mastered the language with the many French workers in the silk house.

When his services were no longer required, he journeyed to Spain in ill health, but, thanks to the help of a kindly German expatriate, he found a position as a deckhand onboard a Spanish ship transporting troops to Cuba to put down one of the frequent rebellions in that colony. In Cuba he was befriended by the U.S. consul, whose hospitality Krueger enjoyed for several weeks before the American advised him to leave Cuba for New Orleans, where opportunities for young men abounded. In New Orleans, Krueger heard of Texas and its limitless cheap land, and the young German became hopelessly struck with Texas fever.

Krueger landed in Galveston and soon found employment at a ranch as a cowboy, where he learned from Hispanic vaqueros how to ride, rope, and handle cattle. After the spring roundup, his services would not be needed until the next year, so he began making his way further west. He found employment with a rancher near San Saba, Texas, running a lumber and flour mill. Krueger relates in his published memoirs numerous exploits of hunting and fighting with raiding Comanches, while also giving the background of German settlement in Texas. Eventually, Krueger was called back to Germany to collect an inheritance. In his native land, he took the money he received and invested it in a complete photography kit, which he intended to take with him back to Texas. The frontier state had not seen this new technology, and Krueger thought he could make a pretty profit in this line of work. After a short stint as a photographer, he sold his camera and lenses and invested in a store south of modern-day Blanco, Texas. Here he also became the postmaster and judge for the local court. He married and expanded his economic enterprises to cotton ginning and ranching as he purchased a sizable tract of land near the Twin Sisters, two prominent mountains in the vicinity. These were his halcyon days: his family grew to five sons and three daughters, and he rose in status to become a prominent member of the community.

The prolonged drought of the 1890s and the coexisting economic crisis of 1893 took their toll on his resources. The spring-fed Guadalupe River was reduced to a "string of water holes" in 1894, and these were completely dry by 1896.[27] Krueger recounts the efforts of Dyrenforth to create rain in San Antonio using explosives, and another experiment in which Dyrenforth tried "to force cold air into the heated strata by steam-driven fans arranged in a tower built for this particular purpose."[28] This later attempt also failed, and the dominant high-pressure system remained unmoved by the suffering of man or beast.

Krueger watched his herds dwindle as weakened cattle lay down in the shade of trees and died. Kreuger reports "deep clefts and fissures six feet deep" opening up in the sunbaked ground, making galloping a horse a very dangerous enterprise.[29] Even though these events occurred twenty-six years prior to the composition of his memoirs, the reader can feel the anguish of the author from passages like the following: "During such times and amidst such surroundings man becomes indifferent to any misfortune that may befall him, as it is beyond his power to comprehend why the forces of nature destroy in a few short years what they have helped to create."[30] The buzzards gathered and began feeding on the animals even before their final breaths, tearing the eyes from the sockets of the dying cattle, and Krueger lamented, "I was never able to understand why thousands of helpless creatures have to suffer so terribly before their final dissolution."[31] There was no way to recover even a small portion of the animals' worth, as their hides had fallen in price so as not to be profitable to process. In 1896 Krueger took a gamble and went into debt to purchase hay from Kansas, hauling it all the way to his ranch to keep his prize herd of Red Poll registered cattle alive, but this effort failed as well. In debt beyond his ability to pay, Max sold his ranch, gins, mills, and what remained of his herd for a pittance of its worth and rode away to New Braunfels with his family. On the day of his departure, he took a few hours to ride to the Twin Sisters and take a last look at what was once his beautiful estate, which took him years to build and where he had spent "the best energies as well as the happiest days of [his] life."[32]

At age 48, Krueger worried that he no longer possessed the vitality to begin again. He took a job as an insurance agent, even though it was a despised profession. Eventually, Krueger recovered financially and opened an agricultural parts business that soon thrived, but he still recalled his agonizing experience with drought. Thousands of small famers suffered the same misery as the drought years of the 1890s slowly succeeded each other.

On the national level, frustrations with the gold standard, railroad monopolies, indebtedness, and the ineffectiveness of the two political parties forced the now thousands of Farmer's Alliance members to look beyond traditional avenues for a remedy to their dilemma. The only means of accomplishing this was to organize collectively, and this the Alliances did when they created the Peoples' or Populist Party in 1891 and nominated a presidential candidate named James Weaver. Others fell behind Jacob Coxey, who urged the unemployed to march on Washington, D.C., to call for works projects to provide government jobs for the 20 percent of Americans who had lost theirs in the throes of the economic crisis. Coxey left from Masillon, Ohio, and other "brigades" of his army departed from sites as widespread as Tacoma, Washington, and Los Angeles, California. As members of Coxey's Army made their way to Washington, D.C., often by jumping on trains as they left the city limits, railroads hired bulls or security guards to check the cars and detain those who "borrowed" the train without paying for a ticket. By the time Coxey made it to the capital, there were close to five hundred thousand followers with him and his son, Legal Tender Coxey. The choice of name for his youngest male descendant need not go unnoticed here, as Coxey's enthusiasm for monetary reform knew no bounds. The "General" was arrested for walking on the grass, and the march dissolved with his incarceration.[33]

Another concern during this era of unbridled industrialism came from critics who claimed factory work was turning assembly line workers into machines. Frederick Winslow Taylor, the son of Quaker parents from Philadelphia, Pennsylvania, was a mechanical engineer who wrote a highly influential book on how to "scientifically manage" factories to increase their efficiency and reduce the costs of production. One aspect of Taylorism, as his ideology came to be called, was to establish daily quotas for factory workers. In most companies, failure to meet the quota would be punished with a reduction in pay, while exceeding the quota would earn the worker a new, higher standard of production to meet each workday. Those who understood that humans had good days and bad days, and that such a system was unfair, claimed that factories were turning humans into machines.

Then there was the emergence of a silver-tongued young politician from Nebraska, William Jennings Bryan, whose speeches as long and shallow as the plains river system had earned the nickname the "Boy Orator from the Platte." At the Democratic Convention of 1896, he stood the assembly on its head with his "Cross of Gold" speech. In the closing moments of his oratory, he assumed the posture of Christ on the cross and thunderously boomed, "You shall not press down upon the brows of labor this crown of thorns, you shall not crucify mankind on a cross of gold!" The reaction was mayhem, as the Democratic Party had not planned to support "Free Silver" or the reintroduction of silver into the currency. Now, Populists had some serious thinking to do, as Bryan became the nominee for the Democratic Party. Should they vote for the third party candidate, James Weaver, and possibly lose the election, or vote for Bryan, and perhaps create an overpowering force by merging their Alliances with the Democrats to win the election? Alliance members ditched the Populist Party and voted in droves for Bryan, thinking the Democrats could win the factory worker vote in Pennsylvania, Ohio, and other industrial centers. Unfortunately for the farmers, Bryan failed to win the vote in any of those pivotal industrial states. Factory owners warned that a Bryan victory would lead to a deeper financial crisis, forcing them to lay off more workers. In a time of viva voce voting, which required people to voice their vote or place colored cards in the ballot box, factory foremen could report on the voting of workers. With the implied threat of unemployment looming over them, most industrial laborers chose to stay home during the election.

All the optimistic activism evaporated like a shallow southern plains playa in late August when Bryan lost the election of 1896. L. Frank Baum published his classic children's tale *The Wizard of Oz* just a few years later, in 1900, and the novel is full of literary coincidences with Populist-era politics. Dorothy, a plains girl from Kansas reminiscent of Mary Lease, teams up with the Tin Man, a bewitched lumberman who only wants a heart to make him human, a mechanical laborer no longer. That allegory to the factory machine-worker argument would have been obvious to readers at the turn of the century. The two are frightened on the road by the Cowardly Lion, whose bark, it turns out, is much more fearsome than his bite—like William Jennings Bryan, who could give the most eloquent speech but not win an election. Dorothy finds the Scarecrow scattered in a cornfield, and after she quickly gathers him together—similar to the Farmers' Alliances building a coalition of farmers—the unlikely companions make their way in Coxey's Army fashion to the capital of Oz. In a land whose name is coincidentally also the abbreviation for ounce, the measurement of mass for precious metals, the quartet travels the Yellow Brick Road—think of the gold standard—to the Emerald City. There they meet the Wizard of Oz, a mysterious ruler who beguiles citizens by manipulating levers on an impressive steam-belching machine, but they find that the wizard cannot help any of them with their problems. He offers magical silver slippers (ruby only in the movie), to whisk Dorothy back to Kansas in a hot air balloon. But she can only return after the heroes vanquish the evil Wicked Witch of the West with a weapon every farmer and speculator wished to exact on the drought-stricken plains: water.[34]

THIRTEEN

And the Skies Are Not Cloudy All Day

Not during the whole trip do we recall seeing a single cloud that suggested rain.

Clyde Muchmore, run participant

Making a home on the range was not as easy a task in the 1890s as the song might imply. Still, an optimistic attitude dominated American minds. In Chicago, the Columbian Exposition opened in May 1893, displaying cultural and technological triumphs to show the world just how far the continent had progressed since Columbus's first landfall 401 years earlier. The Anthropological Building contained exhibits demonstrating the "stages of development" that the indigenous population had undergone since the arrival of Europeans.[1] On the same 664-acre fairground, the Midway Plaisance and the Exposition buildings were illuminated by marvels of new technology: 120,000 incandescent bulbs and 7,000 sputtering arc lamps. Exhibits from 46 states and 36 countries documented the contributions each had made to civilization and seemed to point the way to a bright and glorious future for the emerging nation.[2]

Chicago also hosted the American Historical Association conference at the exposition, honoring the organization's tenth anniversary. In a conference presentation entitled "The Significance of the Frontier in American History," a thirty-three-year-old sandy-haired historian named Frederick Jackson Turner claimed that the most influential factor in shaping the character of the United States had been the frontier, and it now was closed.

Turner pointed out that, according to 1890 census data, which defined a "frontier" as a county in which there where fewer than two residents per square mile, the era of the frontier in American history had passed. This was a cause for concern; for Turner, it was the frontier that had made the United States strong, weeding out all but the most energetic and industrious and promoting a democracy in which men were judged not by their ancestry but by their competence. According to many proud U.S. citizens of the time, their country had risen to prominence on the backs of those stalwart pioneers who tamed a continent. As hundreds of thousands of people immigrated to the United States from eastern and southern Europe, where they had no contact with the purgative experience of the frontier, the nation could lose the elemental ingredient that had ushered it to success.[3]

It appeared as if the United States would soon collide with the perceived societal decay that had hit the major European powers: a rising lower class population and an increasingly effete aristocracy. The United States needed a safety valve to let off some of the population pressure of the crowded northeastern cities. Although the days of the frontier had passed, the nation still held some open land on which to settle working-class families, allowing them to develop the virtues of hard work, self-reliance, and productivity that had inspired the nation's forefathers.

A little-known region called the Cherokee Outlet in the nation's interior grasslands could help solve the problems of the United States as it continued to rise as an industrial power. Certainly, less than two people per square mile inhabited these six and a half million acres in northwestern Oklahoma Territory, but not for long. On August 19, 1893, President Grover Cleveland announced the opening of the outlet, which ultimately became the largest area to be settled by a land run. According to a report of the House of Representatives, the government dispensed the land to the public in a manner that would ensure "that the honest homeseeker, though humble and poor, might acquire a good home for himself and his family for a small sum."[4] This idealism continued to influence interpretations of the run, which have diverged into two basic camps: those who celebrate it and those who focus on its less-than-admirable qualities, but none of the investigations of the Cherokee Outlet Land Run have dealt with the impact of the drought on run participants.

However, contrary to popular perception, the Cherokee Outlet Land Run was anything but an opportunity for the humble and poor home seeker. As early as March 1893, seven thousand families had gathered on the northern edge of the outlet in anticipation of its opening to settlement. Due to delays in the run's start date, many land seekers abandoned their squatter camps to seek lands on which they could settle immediately. A newspaper reporter visited ten or so of the squatter camps and found that not one boomer admitted to having enough money or supplies to last until the fall.[5] How then could they last through the winter and spring, waiting for their crops to ripen?

Perhaps more damaging to homesteaders in the Cherokee Outlet was the drought, which lasted from 1893 to 1897. Those waiting to make the run endured incredibly high temperatures all through the summer of 1893 and well into the fall. Procuring water was a constant struggle for both humans and stock. Given these conditions, it is not surprising that one out of every four who held a claim in 1894 was gone by 1897. Drought and its attendant disasters—fires and floods—continued to chase settlers off their claims until the drought ended. Although Congress had opened the outlet to settlement, the environment had not yet given its endorsement.

The Cherokee Outlet had existed for more than sixty years. Prior to their removal to Indian Territory, the Western Cherokees had lived in modern-day Arkansas along the White River and in other parts of the region, until non-Indian settlers encroached on tribal lands in such numbers that tribal elders feared their settlements would again be surrounded. The non-Indian settlers would surely soon demand that the Cherokees give up their land. The Cherokees were also concerned that if the strip were open to settlement, they would no longer be able to continue their necessary hunting forays into the western plains.

A delegation of Cherokee headmen visited Washington to protest these settlements. The trip to the nation's capital resulted in a new treaty and another exchange of land. Cherokees agreed to give up their claims to White River country in 1827 for land in modern-day Adair, Cherokee, and Sequoyah Counties. At that time the area was known as Lovely County, named after William Lovely, the man who negotiated with the Osages for the acquisition of that territory for the United States. The new treaty also established an outlet to the western hunting grounds that was fifty-eight miles wide and spanned four lines of longitude between the ninety-sixth meridian and the one hundredth at the Mexican border. In 1835 the controversial Treaty of New Echota gave the rest of the Eastern Cherokee homeland to the United States, forcing the bulk of the Eastern Cherokees to relocate to territory acquired by the Western branch of their tribe.

As railroads were built across western Kansas, cattle drives from Texas crossed the outlet on their way to train junctions in Kansas. During the late 1860s and 1870s, trail bosses began to see the advantage of lingering in the outlet, grazing their herds on the luxuriant growth of prairie grass that prospered since the bison herds had been hunted to near extinction. The absence of buffalo also precluded the Cherokees' use of the outlet as a hunting range. Still, many members of the tribe saw an opportunity and charged the cattlemen for grazing rights. Initially, the tribe charged $0.42 a head for this privilege, and by 1882 this arrangement netted them a little more than $41,000.[6] In 1880 those cattlemen who owned grazing rights from the Cherokee tribe met at Caldwell, Kansas, to discuss forming an organization that would provide a protocol for settling disagreements among cattle companies and establish a system of law and order in the outlet. This organization came to be known as the Cherokee Strip Livestock Association.[7]

Regardless of the Cherokee tribe's attempts to ensure that it would never again be surrounded by American settlers, the Treaty of Guadalupe Hidalgo in 1848 ended the Mexican-American War and added the area west of the one hundredth meridian to the United States. By the 1880s not only were the Cherokees encircled by other American settlers, but boomers were working to loose the outlet from tribal control.

Powerful special interest groups joined in calling for the opening of the Cherokee Outlet to public settlement. The St. Louis–San Francisco Railway bought five thousand shares in the Indian Territory Colonization Company, headed by the persistent David Payne.[8] Railroad companies employed florid writers such as Elias C. Boudinot and Major Gordon "Pawnee Bill" Lillie to fill the booster newspapers with articles demanding that the outlet be opened. Within the bureaucracy of the Bureau of Indian Affairs there was a growing interest in assimilation. When the Dawes Act passed through Congress and established the process of dividing reservations into individual allotments for tribal members, it created "surplus" land that might be opened to non-Indian settlement. Boomers called for the immediate opening of these regions.

The land run on Unassigned Lands in 1889 exposed some problems with this system of land acquisition, not least of which were keeping settlers from sneaking into the territory before the appointed hour and maintaining law and order in a large region populated by thousands of opportunists. Still, opening one section of Indian lands to settlement created a fever for more. The government opened the Sac and Fox, Pottawatomie, Shawnee, and Iowa lands in 1891 and the following year, another three and a half million acres of "surplus" Cheyenne-Arapaho land. As these land runs occurred, thousands of down-on-their-luck farmers joined with the railroads to pressure elected officials to reconsider treaty obligations to the Cherokees. Those arguing in favor of opening the region to settlement claimed that there were no longer any buffalo in the outlet, and even if there were, the vast majority of Cherokees no longer used the hunt to procure meat. Proponents also pointed out that the leasing of tribal lands to the Livestock Association only ensured revenue for a few Cherokees, not for the whole tribe.[9]

In 1890 President Benjamin Harrison ordered the removal of all cattle from the outlet, thus subtracting an important source of tribal revenue and forcing the Cherokees, Pawnees, and Tonkawas to consider selling their lands to the United States. By the next year, the Cherokees, fearing that they would lose the outlet anyway and receive nothing for it, agreed to sell the strip to the United States for $1.29 an acre. Congress did not ratify this agreement until April 1893, and President Cleveland waited until August to proclaim that the outlet would be opened on September 16 that year.[10]

The date set for the Cherokee Outlet Land Run seemed onerously late to those who had gathered early for the run, even though there were certainly plenty of reasons to delay the opening until the fall. Officials needed to survey the outlet into 160-acre claims and identify the borders so that the run participants would know exactly what quarter and section they were claiming. President Cleveland further wanted surveyors to divide the area into counties with predetermined county seats and workers to dig wells in case there were no other sources of water in the early days after the run. Cleveland gave the responsibility for organizing the run to Secretary of the Interior Hoke Smith. By September 11 all the wells had been completed with "great difficulty," and officials worried over the capacity of the wells at Pond Creek, Alva, and Woodward.[11]

In actuality, the run progressed rather quickly, but for those who had gathered at the Kansas–Indian Territory border, it could not occur soon enough. Daily, their funds and supplies dwindled. Furthermore, the delay until September assured that after the run, they could not get a crop sown and harvested before the coming winter, nor could they grow any garden crops or preserve vegetables for the looming cold months. Such a late start for the run meant that only those with enough resources to make it through the winter would last long enough to keep their claims, since ultimate ownership required six months' residency on the plot.[12]

Not overly worried with the timing of the event, Smith concentrated on avoiding the problems associated with previous land runs. He called on eight troops of cavalry and four units of infantry stationed at Forts Reno and Supply to patrol the outlet and escort sooners, those who illegally entered early, out. Although military units began these missions in July, most of the troops did not arrive in the outlet until August.[13] These soldiers, also responsible for ensuring that the members of the Livestock Association removed all of their cattle from the area, soon discovered another problem: neighboring farmers and ranchers often traveled into the Outlet to cut and bind hay for stock feed. In an effort to force out those already attempting to settle the outlet, discourage other sooners from entering early, prevent farmers from illegally gathering prairie hay, stop ranchers from grazing their livestock in the public domain, and expose the quarter and section markers for the run, the military set fire to the outlet. Though implemented with the best of intentions, the fire added to the suffering of legal settlers. Wind-borne ash plagued registrants during the weeks prior to the run. Afterwards, many claimants did not have pasture to graze their stock as a result of these burns, adding to the difficulty of maintaining residence on the claims.[14]

Smith attempted to foil others who hoped to profit from the land run. Many of those who owned land adjacent to the strip planned to sell the right to camp next to the Cherokee Outlet boundary to run participants and make a handsome sum with little labor. The secretary checked this plan by ordering a strip of land one hundred feet wide to be set aside for settlers just inside both the northern and southern boundaries of the outlet. In some sections on the northern boundary, this strip was plowed so that there would be no mistake as to how far onto the outlet a run participant could camp.[15]

Smith also established registration booths at nine locations along the northern and southern borders of the outlet, including five sites in Kansas, in Arkansas City, Hunnewell, Caldwell, Cameron, and Kiowa; and four sites south of the outlet, in Stillwater, Orlando, Hennessey, and Goodwin. Those wishing to participate in the run would have to register first, swearing that they had not entered the outlet prior to the legally established hour of 12:00 noon on September 16. They were required to present their registration forms after the run when they filed their claims. This strategy failed to end the fraud associated with land runs, for it was all too easy to obtain a counterfeit registration form.[16] The system also required participants to arrive days before the actual event, necessitating a further outlay of money and preventing those with limited funds from making the run.

It is no surprise that, given the opportunism of the run, Smith was unable to preempt fraud. The secretary of the interior caught wind of a scheme involving some Cherokees and the Rock Island Railroad. Tribal members were allowed to take their allotments prior to the run, and a few of these men intended to choose lands in the proposed county seats and sell them to Rock Island for a nice profit. Of course, the railroad company would make out fabulously as well. To counter this activity, Smith ordered new county seats established at least three miles from the original sites. The railroads refused to recognize these new county seats and, even after the run, skirmishes erupted over which sites were legitimate.[17]

The registration booths opened for the first time at 7:00 A.M. on Monday, September 12, and the lines had been forming since 8:00 the previous morning.[18] Staggering numbers of people surpassed the accomodations of the modest towns. The *Guthrie Daily Leader* reported on September 12 that registrants in line numbered 3,000 at Stillwater; 4,900 at Hennessey; 5,000 at Orlando; 8,000 at Kiowa; 10,000 at Caldwell; and between 8,000 and 10,000 at Hunnewell, Cameron, and Goodwin. The largest gathering of registrants was at Arkansas City, where 15,000 lined up at the booth that had been set up outside city limits on the open prairie, four miles from water, shade, or latrines. In fact, town leaders were so concerned over a recent shortage of rainfall and its effect on their water supply that they callously forbade registrants the use of any city water.[19]

At Cameron, water problems were just as serious, yet the authorities chose to err on the side of humanity: those awaiting the run just south of town were allowed to enter the outlet in search of water, but they were permitted this convenience only under the most rigid restrictions of traveling straight to the source and returning immediately.[20] Yet some registration communities were quite festive in anticipation of the run. Ad Moore, whose family camped at Cameron, claimed, "It was quite a two days wait with two saloons, a carnival, and revival meetings all going on at the same time."[21]

Even with these attempts to preoccupy the participants with festivities, evidence of drought was everywhere. All through the summer of 1893, rain had failed to moisten the earth. The *Beaver Advocate,* printed in the panhandle of Oklahoma Territory, published the U.S. Department of Agriculture's weather bulletin, which stated that "rain is badly needed in the Cherokee outlet, on its borders are encamped many thousands awaiting the opening, all small streams are dry and the crowds have to go many miles for water."[22] John T. Meese recalled that "hot winds had been blowing for months and all the vegetation was dry as tinder, and the earth was cracked open till you could run your hand down."[23] The weather took its toll on animals as well. More than two hundred horses perished on the road between Guthrie and the nearest registration site of Orlando during the week prior to the run.[24] Secretary Smith ordered railroads to haul carloads of water to towns in the outlet in the hope of alleviating the suffering that was sure to follow the excitement of the run.[25]

As a result of the heat and the drought, many spontaneous business ventures opened. In most registration sites, water was selling for anywhere between a nickel and fifty cents a cup and one dollar a barrel.[26] The horse trade thrived, as settlers' horses played out under the strain of carrying their owners to the jump-off sites. From Monday to Thursday of the week prior to the run, 390 horses were sold, and newspapers pronounced that "fresh droves of animals are driven in every day and the prospect is that sales will continue until Saturday."[27] The entrepreneurial spirit was visible in the whole process. Although promising only a modest profit, there were those who held places in line with the intent of selling them for a fee of ten dollars.[28]

The illegal sooners faced their own set of problems. Since the fires had burned a lot of the grass, those who risked unlawful entry into the outlet had to search constantly for forage and water. The military authorities understood the sooners' dilemma, and using a strategy that had proved successful against the Apaches, they waited for illegal settlers at water holes. The *Guthrie Daily Leader* printed a piece titled "How to Catch Sooners" whose author claimed, "It is believed that the surest and most effective way of catching sooners is to watch the few sources of water supply in the territory. Nearly all captured so far have been caught this way. After getting in they find subsistence impossible either for themselves or their animals and are compelled to hunt water."[29]

Meanwhile, the conditions at the registration booths worsened. The dry grass that had not been previously burned off was quick to ignite, and numerous fires crossed the outlet. Some of these were caused by sparks from railroads, and others were accidentally set off by settlers starting a campfire or clumsily bumping over a lantern. The famous prairie winds drove the ash from these fires into the lines of registrants, stinging their eyes and filling their nostrils, but they could not seek cover for fear of losing their places in the queue. The heat continued and was abnormally high for September. The day just before the run, September 15, was the hottest day on record, with the temperature reaching 108 degrees Fahrenheit in Arkansas City.[30] An accompanying strong wind seemed to suck the moisture right out of the registrants' skin.

The health of those enduring the long lines and the intense heat became an issue. People desperately began drinking from any source of water, even dipping their hands or canteens into stagnant pools that were sure to hold parasites. Some ate food that had sat in hot wagons too long to be consumed without repercussions. Dysentery swept the registration sites, dehydrating those who could not afford the cost of water and forcing many to drop out of line, losing their chance to register and participate in the historic event. Women were especially prone to suffering in the heat under their heavy Victorian-era clothing.[31] The incredibly high temperatures caused a reported fifty cases of heat stroke and at least ten deaths. One of the casualties of the September heat was a Civil War veteran named Mr. Billings, who, proudly wearing his Union military uniform, perished within a few feet of the registration booth.[32]

The day of the run offered no relief from the drought and intense heat. Because of the burns, even the slightest movement across the bare soil produced clouds of ash. The event itself was hardly forgettable, but the harsh elements ensured that September 16, 1893, was a day no run participant would ever forget. Lucille Gilstrap recalled that day years later: "Rain had been scarce in Oklahoma Territory and it was both hot and dusty. . . . The dust was so thick you could hardly breathe and everybody was hot and tired, crowding and pushing, and some fighting and cursing. . . . Some people had been camped there for days and there was no water to drink, except what was available in the Cimarron River."[33] Onlookers, some of whom had traveled miles to see the event, must have been disappointed by the short amount of time they were able to witness it. After the officer in charge of initiating the run fired the starting shots, Joseph Redfern claimed that dust had obscured his view by the time the participants had gone only a quarter of a mile. The hours he spent traveling to the border to observe the land run must have seemed hardly worth the effort.[34] Once again, the environment had wreaked havoc on the best-laid plans of men.

The animals' distress was in many respects worse than that of their owners. Many of the participants' horses were malnourished and suffering from dehydration prior to the run, and their owners were so consumed with claiming good pieces of land that they neglected to take the interests of their animals into consideration, or, if they did, they heedlessly jeopardized the animals' lives. Beneath the spurs of their riders, countless animals were mercilessly driven to their deaths.

The burning of the prairie also meant that there would be little forage on most claims for months after the run, creating even greater competition for claims that somehow still had groundcover.[35] Etta Stocking, who had journeyed from Cripple Creek, Colorado, to brave the conditions in Arkansas City, recalled the land run years later. After the gun was fired into the air, she drove her pony, Billie, ahead of the crowd to an area where "grass was still standing" and staked her claim, then quickly unsaddled her horse and let it graze.[36] Amos Kealiher bragged that his father had claimed the "only land with vegetation" for miles around Helena, Oklahoma. Interestingly, his family's oral history blamed sooners, not the military, for starting the fires in the outlet to discourage others from entering the region.[37] Participants who registered at Caldwell, Kansas, mentioned the effects of fires in Grant County, Oklahoma, revealing the high dispersal of burns, whether intentional and unintentional.[38]

In retrospect, it can be seen that the competition was extremely high for any claim. Of the 6,500,000 acres in the Cherokee Outlet, the U.S. government withheld 735,000 acres from settlement for the establishment of public schools and universities 8,640 acres for the Chilocco Indian School; 5,600 acres for Indian allotments; and 2,500 acres for county seats. This left 5,748,260 acres for settlement, enough land for roughly 36,000 claims of 160 acres each—but there were 110,000 registrants.[39] That means that slightly more than two-thirds of those participating in the run were not assured a claim. Furthermore, the great majority of participants registered in the eastern half of the available land, making competition even fiercer there.[40] The *Guthrie Daily Leader* reported that every quarter section of quality land in the vicinity had at least two persons contesting the claim, and some sections had five people vying for ownership.[41] The Cherokee Outlet Land Run, advertised as a grand moment of opportunity, would end in dismal failure for most.

To make matters worse, drought had dominated the region for months. Even before the run, editorials wondered what settlers would do when they arrived. One predicted: "the cry will be water and not whiskey on the Strip. Water [will be bought] at any price and of any quality." Another quipped, "when they get to the Strip, what will the boys do with the canteens? There is no water there to fill them."[42] On September 24 the *Guthrie Daily Leader* stated that no trace of rain had fallen on the eastern portion of the outlet since April 1, and the drought showed no sign of ending: September 24 was dry as well.[43] The U.S. Department of Agriculture echoed this appraisal for western sections of the outlet in a *Beaver Advocate* article on September 28, 1893: "We are now passing through one of the longest sieges of drouth ever experienced, no rain has fallen for twenty-two days and the present prospect for rain is not encouraging."[44] Such conditions baked the soil into a hard crust that would prove extremely difficult to break for planting or for the construction of the most popular improvement, a sod house.

The railroad companies did provide water at town sites in which they had interests, but they boycotted the new county seats of South Enid, Perry, and Pond Creek, forcing residents in these areas to depend on water from local streams or hastily dug wells. Even the efforts of "Rainmaker" Jewell were unsuccessful in coaxing moisture from the skies.[45] A homesteader was lucky indeed to have a neighbor who had access to water and was willing to share it. In many cases, however, free enterprise reigned. Ed Hungerford, either by foresight or blind luck, had a water-producing well near Turkey Creek. In a truly egalitarian spirit, Hungerford charged five cents a head, whether it belonged to an animal or a human, to drink from his oasis.[46]

Other businesses took advantage of the economic opportunities created by the drought. By advertising in local newspapers their methods for overcoming the dry conditions, the Aermotor Company of Chicago cashed in on the deficiency of water by promoting the sale of windmill pumps. Loomis and Wyman, out of Tiffin, Ohio, urged settlers to purchase well drilling machines, stressing that they could dig to any depth, from one hundred to two thousand feet. Kansas City–based Rowell and Chase Machinery offered a free catalogue of well equipment, including "augurs, rock drills, hydraulic and jetting machinery." Perhaps the nod for the most inventive advertisements should go to the marketers of Hood's Cures, who claimed that their product could cure a host of hot weather ailments: hives, boils, pimples, and "other eruptions which disfigure the face." They seemed to hit on most health concerns of average Americans when they further claimed that "in hot weather something is needed to keep up the appetite, assist digestion, and give good healthful sleep. For these purposes, Hood's 'Sarsaparilla' is peculiarly adapted."[47] There were, of course, those advertisers who maintained the purest of optimism in the face of such dire circumstances; Tower's claimed its Fish Brand Slicker was "the best waterproof coat in the world" and continued to run ads as the dry months passed one after another.[48]

The problems facing homesteaders in the Cherokee Outlet forced an exodus almost as incredible as the run. The exodus was at first the result of outward-bound participants who had failed to stake a claim. Then, fires, competition over claims, and drought brought many to the realization that they could not remain on the sunbaked prairies. Five days after the run, hundreds of dust-covered boomers boarded trains departing from outlet stations.[49] It appears that the railroads were able to profit from successful as well as failed run participants. Roads were once again crowded with wagons, this time heading away from the outlet. Frank and Mary Crissup of Elk City, Kansas, recalled seeing all the wagons as they came to participate in the run. On one they remembered quite vividly, there was a painting of a large jackrabbit sitting in a field of green grass. Bold letters just below the picture scrawled "Oklahoma or Bust." The day after the run, the Crissups recalled seeing the same wagon heading north with the old script painted out, "Busted" written in its place.[50]

Platted townsites proved no more successful at retaining settlers than the farmland claims. Many who had hoped for a Ponca townsite lot on which to make their new livelihood left within six days of the run, leaving only two hundred or so residents there. In the words of one disgruntled run participant, "we are going back to Texas, where we have water to use."[51] The town of Perry went through a similar boom and bust. Within days of the run, the population of Perry was close to fifteen thousand. A few months later, the town was reduced to some three hundred souls.[52] It is not surprising that businesses, especially the saloons, were also affected by this population loss. Enid sprouted 51 saloons, and Perry claimed 110 in the immediate days after the run. Some two months later, the number of bars had lowered to 37 in Enid and only 52 in Perry.[53] One might say these two towns had become drier with the drought.

The arid conditions continued for the next three years. In August of 1894, the *Mulhall Chief* reported a temperature of 114 degrees Fahrenheit in the shade at Woodward, Oklahoma Territory. Another article in the same paper discussed the effects of the drought on the corn crop: "An Oklahoma professor says that the hot winds of July 1, 2, and 3 dried out the corn tassels to such an extent that no pollen, or very little, was available to impregnate the silks, which is necessary to the formation of perfect grains and full ears."[54] By September the absence of replenishing rainfall during the previous two years became evident in the region's most dependable water supply, subterranean sources. Once again, wells and springs began to dry up.[55]

As spring returned to the prairies in 1896, conditions were perfect for grass fires. The groundcover was extremely dry and brittle after three years of low moisture. Typical spring winds were quick to spread the effects of any careless spark for miles. On April 15 fires raged west of the town of Perry, consuming several houses and "large quantities" of prairie hay. Local officials arrested a Mr. Jones for disregarding the fire laws and causing one of these blazes, and area residents were so angered by his carelessness that there was talk of lynching the man.[56] Fires in Payne County that autumn destroyed "thousands of acres of pasturage, causing loss of hay, corn and buildings to many farms and fatally burning two persons."[57]

Thus, most settlers who arrived in the outlet with high hopes soon found that even if they were determined to stick it out, they would struggle to find an income. As during any prolonged drought, farming simply was not a viable option for many. Yet claimants were caught between needing to meet the six-month residency requirement to keep their land and simultaneously needing to feed themselves. Outlet families proved extremely creative and resourceful in meeting these demands. John Meese, whose claim was one mile east of Lamont in today's Grant County, took his wife and children back to Belle Plaine, Kansas, immediately after filing. The incredible cost of making the run consumed all of his pecuniary resources, so he returned home to live with family members until he could, in his words, "recuperate financially." He worked for wages of seventy-five cents to a dollar a day, saving up enough to purchase supplies meant to last through the six months they would stay on the claim. Meese continued his strategy of working in Kansas for six months and then returning to his claim in Oklahoma Territory for the remainder of the drought and was able to retain his land.[58]

John Leierer staked his claim on the northern boundary of present-day Major County and quickly returned home to Ulysses, Kansas, to make enough money to pay for his filing fee. He made the trip to Alva to file his claim, then went back to Ulysses for the winter. Leierer spent the first half of the spring on his new property before realizing that he would not make a crop that year due to the drought; he went to work near Oklahoma City as a threshing crew hand. He talked a friend into staying with him on the claim the next spring. The two waited from April 10 to May 10 for rain. When none came, they traveled back to Ulysses to find work for the rest of the year.[59]

Cap Holton kept a job as a car inspector for the Frisco Railroad at Caldwell, Kansas, even though he made the run and held a claim near Cleo Springs in present-day Major County. His wife and children stayed on the property while he worked in Caldwell and came home to visit them whenever the possibility arose. In this way he and his family were able to keep their claim through the drought. He later recalled that "about the time President McKinley was inaugurated," he came home for good "because there were other families near."[60] McKinley assumed office in 1897, the year rains returned to the southern plains and ended the drought.

Others were not fortunate enough to find adequate incomes while the drought reduced crop yields. In the fall of 1893, the Oklahoma Territorial Legislature appropriated ten thousand dollars to purchase seed for needy farmers, though Governor Abraham Jefferson Seay actually spent only $6,460.94 for this purpose.[61] The government also issued rations of bacon and beans to struggling families so that they could make it until they could get a crop planted.[62] Of course, there were no substantial yields for three years. Some received military pensions from their service during the Civil War, whereas others relied on aid sent from family and friends who lived outside the outlet.[63] The *Edmond Sun Democrat* described the conditions of families living in today's southern Grant County:

In consequence of the serious drought of last summer and this spring, there is neither grain, garden vegetables, nor grass for animals. The people in [this] drought stricken section are in utter and deplorable destitution. Many families are now without the common necessities of life, and are compelled to subsist in many cases on cornmeal and water and cracked wheat and water. There are many families which are now without money to procure even the coarse food, and unless help is forthcoming they must face the prospect of starvation. An appeal in behalf of the unfortunate inhabitants of that section has been issued by the Women's Aid Society of North Pond Creek, OK. Food, clothing and garden seeds are solicited.[64]

Reminiscent of earlier attempts to raise relief for those suffering from the drought, a committee of three leading citizens from the small community of Nash in present-day Grant County traveled as far afield as Kansas City, Chicago, and Denver to drum up donations of food and clothing to take back with them. Andrew Anthony of Pond Creek recalled a load of supplies that arrived from Missouri, which a community storekeeper housed and allowed the needy to use free of cost.[65] Kansas Mennonites sent relief to their brethren in the outlet as well.[66]

Unfortunately, these charities were not sufficient for the majority of settlers on the Cherokee Outlet. An investigation of tax rolls reveal just how debilitating the drought was on maintaining land occupancy. Nearly 30 percent of those filing property taxes in 1894 were no longer on the land two years later. Interestingly, women were more prone to stick out the drought and remain on their original claims, as were those with more available wealth. Those with little taxable property were twice as likely to give up their claim as those with sizeable property qualifications.[67]

It is evident that those with greater means were able to treat the run as an investment. If it did not pay off, they could minimize their losses by selling the claim and returning home, but the poor could not maintain their claims as easily and more often would have to quietly abandon it. The prolonged drought affected land evaluations as well. As years passed by without yields great enough to return a profit, claimants had less capital available to invest in improvements on the land. Fifty-one percent lost valuation in their land even though they remained on the same plots.[68]

Even those with the financial resources to make the run took a beating. Ed Bradson owned property in Newkirk Township in K County that tax assessors valued at 468 points in 1894, a relatively high number that reflects a sizeable number of livestock, as it would have been difficult to construct the same value in barns, fences, or corrals in one year's time. But by 1895 Bradson's valuation had dwindled to a mere 38 points. To no surprise, he was not recorded in the 1896 tax roll. J. S. Gilbert of Salt Fork Township in L County provides the most extreme example of falling valuations. His property tallied a valuation of a whopping 2,095 points in 1894. In the 1896 tax roll, his property was valued at 26 points. There were instances in which those of meager means were able to improve their situations. F. H. Nichols of Lowe Township in K County showed amazing resourcefulness by increasing his property valuation from 12 points in 1894 to 25 points two years later. Still, property for the majority of settlers was immediately valued above 30 points. Most of these plots lost value during the first three years as homesteaders sold their possessions to purchase the supplies that would allow them to remain on the claim for the required six months.[69]

In the following year, 1897, rains returned to the region, and property values that year reveal the influence drought had on land assessments in previous years. In the four townships mentioned in O County, 45 percent of the property holders' valuations rose, while 39 percent fell and 15 percent of the claims had been abandoned.[70] Many settlers later recalled 1897 as the year their fortunes changed. J. W. Kephart, whose homestead was five miles east of Carmen in present-day Alfalfa County claimed it was the first year he was able to raise a wheat crop. Charlie Bennett, who resided three and a half miles south of Helena, stated that he was finally able to make a little money from his crops that year, and, as mentioned earlier, it was also the year that Cap Holton decided to resign from his railroad job in Caldwell, Kansas, and live on the claim with his family.[71]

By 1897 the last major drought of the nineteenth century had run its course. In its wake, it had left many families devastated financially. The Cherokee Outlet Land Run did not deliver on all the lofty goals espoused at its conception and ultimately was anything but an opportunity for the "humble and poor" home seeker. If there was any opportunity, it presented itself only to merchants at registration sites, who profited from the thousands of run participants requiring feed for their livestock, flour and other commodities for their families, and lumber, wire, and windmills for the improvement of their claims. The other big winners were the railroad companies, which had lobbied for the opening of the outlet and profited from transporting both eager participants to the jump off points and despondent failures on their way back home. Their glowing advertisements of the plains had done much in promoting the run in the first place.

The Atchison, Topeka & Santa Fe Railroad had produced a pamphlet entitled *Cherokee Strip and Oklahoma: Opening of Cherokee Strip, Kickapoo, Pawnee and Tonkawa Reservations* in 1893, portraying the outlet as truly inviting settlement. Aside from giving specific information on the qualities of the land, guidelines for staking a claim in the run, and ticket information for travelling to registration sites, it claimed that "east of the 98th meridian rainfall is said to be certain"—remember the outlet ran from the ninety-sixth to the one hundredth meridians. The brochure further estimated that from the far northwestern corner to the southeastern tip of the outlet, between twenty-three and thirty-five inches of rain fell annually, though it failed to state which base years had been used to project these estimates. Rumors circulated back east of the dry tendencies of the plains. Perhaps in an effort to dispel this myth, the pamphlet compared the outlet's estimated rainfall totals with points east such as Milwaukee, Wisconsin, and Mackinac, Michigan, which it stated received only thirty and twenty-three inches annually. Then the brochure summed up its understanding of aridity in the region by stating quaintly, "it don't much look like a continued drouth, does it?"[72] Given the seductively optimistic tone of the piece, one suspects that even if AT&SF writers could see into the future, they probably would have printed their pamphlet anyway.

In 1873 Dan Kelly, a musician of Gaylord, Kansas, received a poem from a friend who requested that he put it to music. The author, Dr. Brewster Higley, was a local physician, who had lived on the prairie for a few years and had come to love the natural simplicity of the region. His poem, originally titled "My Western Home," was to become one of the more popular lyrics of the plains. It is known today as "Home on the Range." The lyrics still evoke the heritage of the region: "Oh give me a home where the buffalo roam / Where the deer and the antelope play / Where seldom is heard a discouraging word / And the skies are not cloudy all day." In the summers of 1893 through 1896, these lyrics seemed out of place on the outlet, as buffalo were seldom seen, and even deer and antelope found it difficult to find forage on the burnt grasslands. As we have witnessed, discouraging words were rampant, but it is true that the skies remained free from clouds all day.[73]

Epilogue

*Finally, the interested parties and sponsors were convinced
that the only way to overcome these recurring dry periods
was to utilize subterranean water courses, and much is now
annually done in that respect.*

MAX KRUEGER

Mercifully, the drought that had economically ruined Max Krueger
and thousands of other farmers and ranchers across the southern
plains receded into memory. In his words, "In the autumn of 1896, after
the drought had spent its force, we were blessed with magnificent rains,
and in a few short months all the pastures flourished more beautifully than
ever."[1] The drought had induced migration and brought fires and countless
hardships for residents of the region, just as droughts have done so long as
humans have inhabited the southern plains. No faith in science or implementation of it could alter the amount of rainfall that fell to the earth.

By the late 1890s and early 1900s, misconceptions about climatic conditions in the southern plains were slowly being rectified. A new gospel
replaced the faith in artificial rain as irrigation became the object of many
agriculturalists' hopes. Non-Indians' earliest attempt to irrigate more than
an individual garden in the southern plains occurred along the Arkansas
River in Colorado when George Swink organized a collective effort to construct a ditch at Rocky Ford in 1873. By 1882 the collective was legally incorporated as the Rocky Ford Ditch Company to construct irrigation works.[2]
Between 1883 and 1887, Otis Haskel set aside his Denver real estate interests
and organized the Arkansas River Land, Town and Canal Company. As
the name of his company implies, Haskel was a man of broad interests and

schemes. Before he could accomplish much with his new enterprise, he sold his brainchild in 1897, and it was renamed the La Junta and Lamar Canal Company.[3]

Farther east, modern irrigation still centered around the Arkansas River in Garden City, Kansas. In 1880, four miles west of Garden City, L. H. Armentrout constructed a ditch from which he watered one hundred acres.[4] Armentrout's success encouraged the establishment of three new irrigation companies in the area by 1884: the Garden City Irrigation Company, the Kansas Irrigating and Water Power and Manufacturing Company, and the Minnehaha Irrigating Company of Topeka, Kansas. From 1881 to 1889, different groups constructed 336 miles of canals along the Arkansas River to irrigate 70,000 acres, causing a population explosion along the western Arkansas River. The Kansas portion of the valley held 10,000 residents in 1884, but by 1888 the population had risen to 70,000.[5]

Even so, there were plenty of Kansans who did not have the foresight or wherewithal to obtain land next to the Arkansas River. During the drought of 1893–1896, necessity encouraged farmers to find new methods for obtaining water or withdraw from the region. Windmill and pump companies advertised heavily in the southern plains newspapers, a wise strategy to increase sales as farmers turned to subterranean water supplies for moisture. By the end of 1895, the Kansas Board of Irrigation, Survey, and Experiment reported that out of 1,335 farmers implementing irrigation, 998 utilized wells as their source, and the great majority of those used wind to power their pumps. From 1895 to 1896, the amount of irrigated farmland in Kansas almost doubled, from 11,823 acres to 22,000 acres.[6]

In Texas irrigation was slower to develop. The XIT ranch began using windmills to pump water from the Ogallala Aquifer as early as 1887, but intensive crop irrigation was not possible on the Llano Estacado until newer, more powerful irrigation pumps made their appearance in the region during the early 1900s. Still, irrigators drilled only three hundred wells that utilized this type of technology between 1910 and 1920.[7] The relatively small amount of investment in large-scale irrigation was due to the high demand for beef and wheat in response to the devastation of Europe's agricultural output after World War I; neither cattle ranching nor wheat farming required irrigation. The availability of long-term credit, introduced during the New Deal, allowed irrigation to resurface during the 1930s. High cotton prices, drought, and economic depression convinced many Staked Plains agriculturalists in northwest Texas to turn to irrigation. By the 1950s, there were several million acres under cultivation on the Texas high plains.[8] The federal government played a major role in settling the region.

The severity of the 1893–1896 drought led Frederick Newell, Assistant Hydraulic Engineer of the United States Geological Survey, to declare the high plains a region susceptible to periodic intense drought in 1896. He pointed out that the best way humans could prevail in such a region was through reclamation projects, which would impound water behind massive dams in the major rivers of the West.[9] Other southern plains irrigation advocates agreed with Newell's concept of government-sponsored reclamation, but by the early 1900s, the return of healthy rains convinced most plains residents to withdraw their support of such a policy.[10] Still, with the backing of railroads, progressive reformers, President Theodore Roosevelt, and political boosters from the western mountain and intermontane states, Congress passed the Newlands Reclamation Act in 1902.[11]

Newell's assessment of the high plains is certainly accurate for the whole southern plains in general, as extended droughts struck the region almost every alternate decade through the twentieth century: 1909–1918, 1933–1938, and 1950–1956. The Dust Bowl of the 1930s ranks first in environmental destruction among the twentieth-century droughts, partly as a result of human actions. The high wheat prices following the Great War encouraged both farmers and investors to purchase tractors along with other new implements to break the plains soil. In the early 1920s, booming agricultural prices seemed to justify these investments, but as Europe recovered from the devastation of the war and began producing its own crops, the price of wheat fell. Southern plains farmers often opted to cultivate marginal land in an effort to make enough money to pay back their loans.[12] When successive dry years engulfed the southern plains, the amount of wind erosion increased dramatically.

Agricultural enterprise had not created the dust storms, but it had certainly provided the circumstances to make the storms' effects more severe. In 1933 the Oklahoma panhandle town of Goodwell experienced a severe seventy days of dust storms. In May of 1934, dirt storms carried soil eastward to New York City, Boston, and as far as three hundred miles out into the Atlantic Ocean.[13] The following years continued to produce an incredible number of dirt storms: 22, 53, 73, and 134 storms hit Goodwell during each of the years between 1934 and 1937.[14]

Surprisingly, the less ecologically severe drought of the 1890s was much more damaging socially. In Richard Warrick and Martyn Bowden's report

"Changing Impacts of Droughts on the Great Plains," the two found the population displacement of the 1890s drought to far surpass that of the famous Okie migration of the Dirty Thirties. During the mid-1890s, a majority of southern plains counties experienced more than a 50 percent population loss, whereas only a relatively small area in southwest Kansas and the Oklahoma panhandle had such high attrition rates during the 1930s.[15] Both Warrick and Bowden attribute this apparent contradiction to the activities of the federal government. During the 1890s as well as the dry trend of 1909–1918, federal governmental officials were reluctant to provide relief. As before, opposition to government aid came from the local level, as boosters and ranchers lobbied on a national level to oppose governmental aid to drought victims.

In context with the government's responses to the previous droughts, the New Deal appears to be quite revolutionary. As during previous droughts, state-organized relief continued but was not equal to the task due to the severity of the depression, and the Franklin D. Roosevelt administration implemented the Agricultural Adjustment Act to attempt to limit crop and livestock production in 1933.[16] The New Deal also tried to retire marginal lands from cultivation and overgrazing, removing farmers from the region in various "Resettlement" programs. Finally, Congress passed the Taylor Grazing Act in 1934, ending the seventy-two-year-old policy of making homesteads available from the public domain and, at least theoretically, limiting the number of livestock allowed by contract on federal land.[17] Furthermore, the New Deal introduced ecological changes, planting shelterbelts to a greater degree than had previously been attempted.[18]

New government programs encouraged the use of environmentally sound land management practices such as terracing, soil mulching, deep tilling, strip cropping, contour plowing, and summer fallow. Though most of these practices were aimed at retaining moisture, they obviously helped retard the effects of drought as well. During 1934 the national government provided $500,000,000 of drought relief to the Great Plains states and $14,000,000 more for the establishment of shelterbelts.[19] In the Texas Panhandle, agricultural production rose $37,737,000 from 1935 to 1942, but it cost American taxpayers $43,327,000 in federal aid.[20] All of this governmental relief curbed the social effects of the extended 1930s drought on the southern plains. We will never know how many more people would have been displaced, how much more crime would have occurred, or how much more suffering plains residents would have endured without these programs.

The next significant drought hit the southern plains during the 1950s. The 1950s drought was more severe than the 1930s episode, but this time federal policies were in place to deal with this type of catastrophe.[21] In 1952 eighty-mile-per-hour winds picked up enough dirt to build a dust front twelve thousand feet tall across portions of the southern plains. Once again, a desert-like topography returned as dunes spread over sections of the region. Baca County, Colorado, experienced a 70 percent loss of its 500,000 acres in wheat. Hamilton County, Kansas, likewise lost 95 percent of its crop, and by 1954, wind erosion had damaged 11,700,000 acres.[22] The Agricultural Experiment Station at Garden City, Kansas, reported thirty dust storms from February 19 to June 30.[23] This 1950s drought, although quite severe, caused less population displacement due to a greater willingness by the national government to provide relief. Even though the financially conservative Eisenhower administration held office, it quickly provided up to $750,000,000 worth of aid to the drought area.[24] Governmental subsidies were absolutely mandatory to maintain a high permanent population on the plains.

Since the removal of Native peoples from their tribal homelands, a stationary population has been growing steadily on the southern plains, whereas before the region's population was much more fluid. Before European colonization, the Paleo-Indians settled along the region's river valleys, up out of the flood plains, but near enough to benefit from the timbered riverbanks. Natives fished the rivers and creeks, hunted the wooded valleys, farmed along the rivers and streams, and on occasion ventured onto the uplands to hunt bison. Precontact agricultural people also lived along the southern plains river valleys, and established a seasonal migratory tradition. During the spring, the people planted their crops and tended the fields. In early summer, they ventured out of the valleys to hunt bison, returning to their villages with hides and meat. They harvested their crops during the late summer and fall, and then made a final buffalo hunt while the bison coats were thickening in preparation for the coming winter. With the crops cached, the meat cured, and provisions ensured, the people remained in their villages throughout the cold winter months to begin the process again in the spring.

Spanish reintroduction of the horse to the Americas created the possibility of not only maintaining a livelihood on the high plains, but accumulating wealth in the form of horse herds. As the equines thrived on the grasslands, other tribes ventured onto the plains to take advantage of the growing bison hide trade triggered by Europeans. The Plains nomads often burned off their territorial grasslands during the late winter to attract buffalo herds with the new grasses that would emerge during the spring. If the following spring proved to be dry, these fires could remove the vegetative cover and expose the topsoil to wind erosion. During the summer, the tribes came together in large groups to celebrate traditional festivals and to gather for the big buffalo hunt. As fall approached, the hides were at their best and the hunting continued, but by late fall the tribes split into small bands and moved to the river valleys to take advantage of shelter in the timbered areas. Here, they remained close to the cottonwood trees lining the rivers and creeks, which provided the bark they fed their horses when there was no other fodder.

Plains nomads were part of an intricate trade network. Specializing in procuring hides from the bison herds, they relied on trade to obtain carbohydrates and manufactured goods. As neighboring enemy tribes received guns from their European trading partners to the east, Plains Natives scrambled to create trade relationships that would provide them with such weapons. Thus the tribes' welfare was dependent on the buffalo. When periodic droughts hit the southern plains and the bison herds migrated out of one region, these nomads could simply follow the herds and maintain their lifestyle, but the United States altered this situation when it removed the eastern tribes to the eastern margins of the southern plains. When drought scorched the region, intertribal warfare erupted during the late 1840s, 1850s, and early 1860s as the United States made a weak attempt to secure peace by negotiating treaties with the nomadic tribes of the plains. In these treaties the United States promised to provide annuities and goods to the migratory bands in return for their promise to keep peaceful relations with their neighbors. The federal government also promoted peace and protected the removed eastern tribes by maintaining a military presence in the region. At times the Plains Indian hunters were desperate enough to go in search of the buffalo herds and risk warfare with their eastern Indian neighbors, which would surely provoke a punitive expedition from the United States.

During times of drought, the nomadic tribes of the high plains, the semisedentary agricultural tribes of the prairies, and the removed eastern tribes became more dependent on their annuities. The nomads were unable to follow the bison herds by venturing out of the southern plains, and the agricultural tribes could not rely on their withered crops. As the early days of the Civil War passed, the Lincoln administration refused to grant any but the most miserly relief to the removed Five Tribes, and went so far as to cut off their annuities for fear the money would end up in Confederate, proslavery hands. Beset by successive years of intense drought and increasing attacks from their nomadic western neighbors, all Five Tribes made treaties with the Confederacy, which promised to continue the annuities.

After the war, these treaties provided the excuse the United States needed to justify taking more land from the removed tribes. The government used what it considered excess lands to settle other removed tribes, but soon succumbed to popular demand to open up all of the remaining public lands freed up by tribal allotment. Government actions had made it possible for a large non-Indian population to move to the southern plains and take up intensive agriculture. During the 1870s and 1890s, the population of the region grew dramatically from 23,624 to 515,671.[25] Most of the people came from the more humid East, and their farm methods were not successful until modified. Drought during the mid-1880s and 1890s created famine and starvation in the southern plains, and masses migrated out—except for the Native Americans, who were not free to do so. The Native Americans who forcibly resided on reservations were subsidized by the federal government, or they too would have had to leave the region or perish.

The differences in Indian and non-Indian interaction with the southern plains environment during times of drought are significant. The Native tribes of the high plains did not practice intensive agriculture. Instead, they planted in river valleys and mixed their gardens with the crop triad of beans, squash, and corn, whose respective root systems reached to varying depths to tap the soil's moisture. During times of drought, people migrated to areas of moisture, to the Caprock Escarpment or other sites where springs bubbled up out of the rock, or even further east, if need be. This migration was important for locating sufficient sources of water, but also for following the bison herds and other game that were also searching for the life-giving liquid.

When wet weather returned to the southern plains, the people and animals did likewise. That seasonal migration worked best with low population density in the region. As the population rose during the 1830s with the arrival of the removed eastern tribes, and especially later during the 1880s and 1890s with the arrival of non-Indian settlers by the thousands, the territorial claims of these settlers made migration unfeasible. Non-Indian settlers arrived on the grasslands with farming methods better suited to a more humid climate, and their attempts to cultivate the upland areas of the southern plains met with miserable failure.

The environment has limited the possibilities for subsisting on the southern plains. The non-Indian settlers had to change their traditional methods of farming and their view of government involvement in agriculture in order to remain on the grasslands. Still, they did not react to drought as the Indians had. Non-Indians continued to maintain high population density on the southern plains, though many of them did migrate during times of drought. However, the implementation of irrigation, dry land farming methods, and soil conservation practices has allowed a much higher permanent population in the southern plains, but only with government assistance.

During the twentieth century, technological advances, governmental economic supports, and changes in agricultural thinking allowed the growing population to remain in the region. There is now a price to pay. Irrigation and flood control dams constructed along the Arkansas River and its tributaries have begun to silt up and impede the river's flow, while below the dams, the salinity of the river water has increased.[26] Higher populations of both humans and animals consume a greater amount of water, and irrigation has proven the only way to sustain agriculture on the high plains. These factors have resulted in a dramatic lowering of underground water reservoirs, which impacts the number of acres cultivated. The Ogallala Aquifer dropped fifteen feet from 2005 to 2015, and in Kansas, a state university study found that if the current levels of pumping are continued, "69 percent of the Ogallala Aquifer will be gone in 50 years."[27]

So, at the end of the day, what can a study of past droughts accomplish within the broader study of man's interaction with the environment? Such a study can point out that the only way the United States has been able to maintain permanent residences and livelihoods for sizeable sedentary human populations on the high plains has been through government subsidies. Not only did taxpayers pay for the removal of Native Americans from their widespread domains, but federal dollars also subsidized the railroads constructed through the plains, allowing farmers to participate in commercial agriculture. Military bases further stimulated local economies, and these days states and the federal government subsidize the region's universities as well. Federal subsidies prop up the price for cotton and corn, the two most profitable crops on the Caprock Escarpment. Between 1995 and 2012, the U.S. government pumped $32.9 billion into cotton production.[28]

Without government aid, the price for cotton could not offset the cost of pumping up water from deep below the surface, nor could farmers maintain the machinery to plant and harvest the crop or afford the chemicals used to kill insects and weeds and to defoliate the cotton. With the lowering water level in the Ogallala Aquifer, this method of agriculture is not sustainable. Either the federal government will find a way to provide water to the southern high plains, or the region will generally return to ranching and dry land farming. A host of jobs related to intensive agriculture, namely, irrigation equipment sales and servicing, agricultural implement sales and parts sales, will no longer be needed, and those workers may be forced to move elsewhere. This shift, if it occurs, will also represent a return to extensive land-use practices that resemble the Native people's reliance on the buffalo.

Could Katharine Bates, at her most creative, ever have imagined the plains as they are today? Fields of cotton, sorghum, and corn, watered by center pivot irrigation systems, create giant circles of green that one can see from a window seat aboard any of the many passenger jets that leave white vapor trails across the wide southern plains skies. There are also an increasing number of fields flowing in the late spring with "amber waves of grain," as dry land farming returns with every capped well that signifies the lowering subterranean water table. Katharine Bates wrote her poem during the drought year of 1893, subscribing to the notion popular in her era that rainfall was increasing. She died on March 28, 1929, roughly four months before the stock market crash of Black Thursday, and one year before the most infamous environmental disaster to hit the southern plains in recorded memory. More ominous is the fact that the 1930s drought was not even the most severe to hit the region in the last two centuries. Though the plains are certainly no desert, it is folly for any people to believe they have conquered drought on the southern plains.

Notes

INTRODUCTION

1. Dan Flores, *Horizontal Yellow: Nature and History in the Greater Southern Plains* (Albuquerque: University of New Mexico Press, 1999), x.

CHAPTER 1

1. Martyn Bowden, "The Perception of the Western Interior of the United States, 1800–1870: A Problem in Historical Geosophy," *Proceedings of the Association of American Geographers* 1 (1969): 19. Most of the nation's elite universities were located in New England and would have greatly influenced local views of the plains region.

2. Ralph C. Morris, "The Notion of a Great American Desert East of the Rockies," *Mississippi Valley Historical Review* 13, no. 2 (1927): 193; Francis Prucha, "Indian Removal and the Great American Desert," *Indiana Magazine of History* 59 (December 1963): 302; Richard Dillon, "Stephen Long's Great American Desert," *Proceedings of the American Philosophical Society* 111 (April 1967): 93; Terry Alford, "The West as a Desert in American Thought Prior to Long's 1819–1820 Expedition," *Journal of the West* 8 (October 1969): 516; Bowden, "The Perception of the Western Interior of the United States,"17; Martyn Bowden, "The Great American Desert and the Frontier, 1800–1882: Popular Images of the Plains," in T. K. Hareven, ed., *Anonymous Americans: Exploration in the Nineteenth Century Social History* (Englewood Cliffs, NJ: Prentice Hall, 1971), 51–52.

3. Thomas Say, *Account of an Expedition from Pittsburgh to the Rocky Mountains,* ed. Edwin James, 2 vols. (Philadelphia: Carey and Lea, 1823), 1:192; and Edwin James, *Part 2 of James's Account of S. H. Long's Expedition, 1819–1820,* in *Early Western Travels,* vol. 15, ed. Reuben Gold Thwaites (Cleveland: Authur H. Clark, 1905), 94.

4. James, *James's Account of Long's Expedition,* part 2: 174.

5. Bowden, "The Perceptions of the Western Interior of the United States," 18. See table 1.

6. John C. Allen, "The Garden-Desert Continuum: Competing Views of the Great Plains in the Nineteenth Century," *Great Plains Quarterly* 5 (Fall 1985): 209.

7. For the letter reference, see Meriwether Lewis and William Clark, *Original Journals of the Lewis and Clark Expedition, 1804–1806,* 7 vols., ed. Reuben Gold Thwaites

(New York: Antiquarian Press, 1959), 7: 310. For the description of the Missouri Breaks, see Stephen E. Ambrose, *Undaunted Courage: Meriwether Lewis, Thomas Jefferson and the Opening of the American West* (New York: Simon and Schuster, 1996), 226.

8. Allen, "The Garden-Desert Continuum," 210.

9. Zebulon Pike, *Pike's Dissertation of Louisiana,* in *Journals of Zebulon Montgomery Pike,* vol. 2, ed. Donald Jackson (Norman: University of Oklahoma Press, 1966), 27.

10. Alford, "The West as a Desert," 524n21.

11. Bowden, "The Great American Desert and the Frontier," 70n20.

12. Ibid.

13. Henry M. Brackenridge, *Views of Louisiana; Together with a Journal of a Voyage Up the Missouri River in 1811* (Pittsburgh: Arthur H. Clark, 1814), 27–35; quoted in Alford, "The West as a Desert," 520. According to Alford, Brackenridge modified this version in 1817 by replacing the description calling the region a desert, with the less offensive depiction of "desert-like." Brackenridge still maintained that the area could never be cultivated. Henry M. Brackenridge, *Journal of a Voyage up the Missouri River in Eighteen Hundred and Eleven,* in *Early Western Travels, 1748–1846,* vol. 6, ed. Reuben Gold Thwaites (Cleveland, OH: Arthur H. Clark, 1904), 161.

14. Dumas Malone, ed., *Dictionary of American Biography* (New York: Scribner's Sons, 1946), 11: 380, 9: 576, 16: 401–2.

15. Merlin P. Lawson, *The Climate of the Great American Desert: Reconstruction of the Climate of the Western Interior United States, 1800–1850* (Lincoln: University of Nebraska Press, 1974), 95.

16. Say, *Account of an Expedition,* 1: 138.

17. Dillon, "Long's Great American Desert," 96.

18. James, *James's Account of Long's Expedition,* 2: 192.

19. John R. Bell, *Journal of the Long Expedition,* in *The Far West and the Rockies Historical Series, 1820–1875,* vol. 16, ed. Harlan M. Fuller and LeRoy R. Hafen (Glendale, CA: Arthur H. Clark, 1975), 103.

20. Roger L. Nichols and Patrick L. Halley, *Stephen Long and American Frontier Exploration* (Norman: University of Oklahoma Press, 1995), 117.

21. Ibid., and 110–11.

22. Ibid., 112–13.

23. Ibid.

24. Bell, *Journal of the Long Expedition,* 99.

25. Nichols and Halley, *Stephen Long and American Frontier Exploration,* 118.

26. James, *James's Account of Long's Expedition,* 2: 191–92. These supplies were meant to provide food and bartering material for the expedition, gifts for the Natives to ensure their cooperation, and equipment for scientific research, but included only 450 to 500 pounds of hard biscuits, 150 pounds of parched cornmeal, 150 pounds of salted pork, 25 pounds of coffee, 30 pounds of sugar, and 5 gallons of whiskey. In Bell's account, the supplies taken for trading purposes included 30 pounds of tobacco, 5 pounds of vermillion, 2 pounds of beads, 2 gross of knives, 1 gross of combs, 2 gross of hawk bells, 2 dozen moccasin awls, a dozen scissors, and some trinkets. Native tribes prized the tobacco, which they smoked prior to any council, and the vermillion, which they used to dye garments red.

27. Nichols and Halley, *Stephen Long and American Frontier Exploration,* 119.

28. Bell, *Journal of the Long Expedition,* 192. The instruments included "three traveling and several pocket compasses; one sextant, with a radius of five inches; one snuff

box sextant; one portable horizon with a glass frame and mercurial trough; one and a half pounds of mercury in a case of box wood; two small thermometers; several blank books, portfolios, & etc."

29. Nichols and Halley, *Stephen Long and American Frontier Exploration,* 99–100. The scientific contributions of this expedition fell far short of the nation's expectations. The party, traveling by steamer, was able to progress only six hundred miles in almost three months; in addition, Congress called Long back to Washington, curtailing the expedition.

30. John Moring, *Men With Sand: Great Explorers of the North American West* (Helena, MT: Falcon Publishing, 1998), 94.

31. Nichols and Halley, *Stephen Long and American Frontier Exploration,* 119.

32. Ibid., 123.

33. James, *James's Account of the Long Expedition,* 2: 262.

34. Edwin James, *Part 3 of James's Account of Long's Expedition,* in *Early Western Travels, 1748–1846,* vol. 16, ed. Reuben Gold Thwaites (Cleveland, OH: Arthur H. Clark, 1905), 82.

35. Ibid., 3: 84. Also, though not included in the quote, a "late storm" had pelted the group with hail and forced them to endure a temperature drop from seventy degrees to forty-seven degrees.

36. Ibid., 3: 93.

37. Ibid., 3: 111.

38. Ibid., 3: 94, 99, 130, and 153.

39. Ibid., 3: 148.

40. Lawson, "The Climate of the Great American Desert," 22, 21.

41. Ibid., 24, 26.

42. Ibid., 22.

43. Ibid., 26.

44. Ibid., 94.

45. Lawson, "The Climate of the Great American Desert," 95.

46. H. Harper, "Drought in Central Oklahoma from Annual Rings of Post Oak Trees," *Proceedings of the Oklahoma Academy of Sciences* 41 (1961): 26–27.

47. Merlin P. Lawson and Charles W. Stockton, "Desert Myth and Climate Reality," *Annals of the Association of American Geographers* 71 (December 1981): 531, 535.

48. Ibid., 530.

49. Ibid., 535.

50. Charles Stockton and David Meko, "Drought Recurrence in the Great Plains as Reconstructed from Long-Term Tree-Ring Records," *American Meteorological Society* 22 (January 1983): 23.

51. E. R. Cook, D. M. Meko, and C. W. Stockton, eds., *U.S. Drought Area Index Reconstructions.* (Boulder, CO: NOAA / NGDC Paleoclimatology Program, 1998). Available online at https://www.ncdc.noaa.gov/paleo/pdsiyear.html.

52. Connie Woodhouse and Jonathan Overpeck, "2000 Years of Drought Variability in the Central United States," *Bulletin of the American Meteorological Society* 79 (December 1998): 2703.

53. Edward Cook, David Meko, David Stahle, and Malcolm Cleaveland, "Drought Reconstruction for the Continental United States," *American Meteorological Society* 12 (April 1999): 1145.

54. James, *James's Account of the Long Expedition,* 2: 120.

55. Daniel Muhs and Vance Holliday, "Evidence of Active Dune Sand on the Great Plains in the Nineteenth Century from Accounts of Early Explorers," *Quaternary Research* 43 (1995): 203.

56. James, *James's Account of the Long Expedition,* 2: 122–23.

57. James, *James's Account of the Long Expedition,* 3: 75.

58. Ibid., 125.

59. Ibid., 135.

60. Ibid., 65, 71, and 126.

61. Ibid., 142–43.

62. Ibid.

63. Ibid., 174.

64. Dillon, "Long's Great American Desert," 97.

65. James, *James's Account of Long's Expedition,* 3: 174. For the location of this particular bed of rocks, see George Goodman and Cheryl Lawson, *Retracing Major Stephen Long's 1820 Expedition: The Itinerary and Botany* (Norman: University of Oklahoma Press, 1995), 111.

66. James, *Part 4 of James's Account of Long's Expedition,* in *Early Western Travels, 1748–1846,* vol. 17, ed. Reuben Gold Thwaites (Cleveland, OH: Arthur H. Clark, 1905), 147.

67. Alford, "The West as a Desert," 516; Morris, "The Notion of a Great American Desert," 193.

68. James Malin, *The Grasslands of North America: Prolegomena to its History* (Gloucester, MA: P. Smith, 1967), 82.

69. Noah Webster, *An American Dictionary of the English Language* (New York: S. Converse, 1828; reprint, San Francisco, CA: Foundation for American Christian Education, 1996), n.p.

Chapter 2

1. James, *James's Account of the Long Expedition,* 4: 13–14.

2. Howard Ensign Evans, *The Natural History of the Long Expedition to the Rocky Mountains, 1819–1820* (New York: Oxford University Press, 1997), 218.

3. Maxine Benson, *From Pittsburgh to the Rocky Mountains: Major Stephen Long's Expedition, 1819–1820* (Golden, CO: Fulcrum, 1988), xii.

4. Ibid., xii–xiii.

5. Edward Everett, "Long's Expedition," *North American Review* 16 (April 1823): 242.

6. Jedidiah Morse, *A New Universal Gazetteer, or Geographical Dictionary Accompanied with an Atlas* (New Haven, CT: S. Converse, 1823), 603, 638, and 651.

7. "Major Long's Second Expedition," *North American Review* 21 (July–December 1825): 178–79.

8. Benjamin Silliman, "Expedition of Major Long and Party, to the Rocky Mountains," *American Journal of Science and Arts* 6 (1823): 374.

9. Hezekiah Niles, "American Desert," *Niles' Weekly Register* 35 (September 27, 1828): 70.

10. For an excellent discussion of these maps, see Francis Paul Prucha, "Indian Removal and the Great American Desert," *Indiana Magazine of History* 59 (December 1963): 299–322.

11. Duane Gage, "Oklahoma: A Resettlement Area for Indians," *Chronicles of Oklahoma* 47 (Fall 1969): 283.

12. George M. Ella, *Isaac McCoy: Apostle of the Western Trail* (Springfield, MO: Particular Baptist Press, 2002), 43.

13. George A. Schultz, *An Indian Canaan: Isaac McCoy and the Vision of an Indian State* (Norman: University of Oklahoma Press, 1972), 103.

14. Ibid., 116.

15. Isaac McCoy, *History of Baptist Indian Missions Embracing Remarks on the Former and Present Condition of the Aboriginal Tribes: Their Settlement Within the Indian Territory, and Their Future Prospects* (Washington, DC: William M. Morrison, 1840), 197–98.

16. Isaac McCoy, *Remarks on the Practicability of Indian Reform, Embracing Their Colonization* (New York, 1829), 32.

17. Jedidiah Morse, *Report to the Secretary of War of the United States on Indian Affairs, Comprising a Narrative of a Tour Performed in the Summer of 1820, under a Commission from the President of the United States, for the Purpose of Ascertaining, for the Use of the Government, the Actual State of the Indian Tribes in Our Country* (New Haven: S. Converse, 1822), 255.

18. McCoy, *Remarks on the Practicability of Indian Reform*, 33.

19. Ibid., 32.

20. Ibid., 33.

21. Ibid., 34.

22. McCoy, *History of the Baptist Indian Missions*, 350.

23. Jeremiah Evarts, "Memorial of the Citizens of Massachusetts," in *Cherokee Removal: The "William Penn" Essays and Other Writings*, ed. Francis Paul Prucha (Knoxville: University of Tennessee Press, 1981), 231.

24. Jeremiah Evarts, "Memorial of the Prudential Committee of the American Board of Commissioners for Foreign Missions," in ibid., 297–98.

25. Ibid., 299.

26. Ibid., 300–301.

27. U.S. Senate, *Views of the Cherokees in Relation to Further Cessions of their Lands*, S. Doc. 208, 18th Congress, 1st session, 1824 found in American State Papers: Indian Affairs, 2: 502.

28. U.S. Senate, *Exchange of Lands with the Indians*, January 9, 1817, American State Papers: Indian Affairs, 2, 145: 124.

29. *Gales and Seaton's Register of Debates in Congress, 1824–1837* (Washington, DC: Government Printing Office, 1825–1839), 3: 75.

30. McCoy, *Remarks on the Practicability of Indian Reform*, 32.

31. U.S. Senate, *A Plan for Removing the Several Indian Tribes West of the Mississippi River*, S. Doc. 218, 8th cong., 2nd sess., 1825, American State Papers: Indian Affairs, 2: 543.

32. Ibid., 544.

33. U.S. Senate, *Report of the Committee on Indian Affairs*, S. Doc. 246, 24th Cong., 1st sess., 1836, 4.

34. Ibid., 4 and 5.

35. Thurman Wilkins, *Cherokee Tragedy: The Story of the Ridge Family and the Decimation of a People* (New York: MacMillan, 1970), 321.

36. Ibid., 36ff.

CHAPTER 3

1. Walter Prescott Webb, *The Great Plains* (Dallas: Ginn, 1931), 153.

2. Josiah Gregg, *Commerce of the Prairies,* ed. Max L. Moorhead (Norman: University of Oklahoma Press, 1954), xviii.

3. Ibid., 49–50.

4. Ibid., 100.

5. Ibid., 235.

6. James, *James's Account of the Long Expedition,* 3: 174; Josiah Gregg, *Commerce of the Prairies,* 356 and 357.

7. Ibid., 356.

8. Ibid., xxx–xxxiii.

9. Ibid., 362.

10. William Morton Payne, *Leading American Essayists* (New York: Books for Libraries, 1910, reprint 1968), 48.

11. Ibid., 68.

12. Ibid., 78–83.

13. Ibid., 85.

14. John Francis McDermott, *The Western Journals of Washington Irving* (Norman: University of Oklahoma Press, 1966), 4.

15. Ibid., 9.

16. Ibid., 84–85.

17. Washington Irving, *A Tour on the Prairies* (Paris: A. and W. Galignani, 1835); Charles Latrobe, *Rambler in North America, 1832–1833* (New York: Johnson Reprint Corp., 1970); and Henry L. Ellsworth, *Washington Irving on the Prairie, or, A Narrative of a Tour of the Southwest in the Year 1832* (New York: American Book Company, 1937).

18. Ellsworth, *Washington Irving on the Prairie,* 156–57.

19. Ibid., 208.

20. Ibid., 57.

21. Ibid., 203.

22. Ellsworth, *Washington Irving on the Prairies,* 46.

23. Irving, *Tour on the Prairies,* 204 and 209.

24. Ibid., 68, 61, 191, and 145, respectively.

25. Ibid., 51.

26. Frederick W. Rathjen, *Texas Panhandle Frontier* (Lubbock: Texas Tech University Press, 1973; reprint, 1998), 88–89.

27. H. Bailey Carroll, ed., "The Journal of Lieutenant J. W. Abert from Bent's Fort to Saint Louis in 1845," *Panhandle-Plains Historical Review* 14 (1941): 18.

28. Ibid., 26.

29. Ibid., 27.

30. Ibid., 57, 61, and 61, respectively, with another reference to desert on page 74 when discussing the *llano estacado.*

31. Ibid., 39 and 32, respectively.

32. Ibid., 79.

33. Ibid., 94, 95. For 1845 moisture estimations see *U.S. Drought Area Index Reconstructions* at https://www.ncdc.noaa.gov/paleo/pdsiyear.html.

34. Carroll, "Journal of J. W. Abert," 95.

35. Ibid., xxii–xxviii.

36. Thomas Hart Benton, *Discourse of Mr. Benton of Missouri before the Mercantile Library Association . . . Delivered in Tremont Temple at Boston, December 20, 1854* (Washington: J. T. and L. Towers, 1854), 8.

37. William Gilpin, *Mission of the North American People: Geographical, Social, and Political,* 2nd. ed. (Philadelphia: J. B. and Lippincott, 1874), 63.

38. David Emmons, *Garden in the Grasslands: Boomer Literature of the Central Plains* (Lincoln: University of Nebraska Press, 1971), 12–13.

39. Ibid., 13.

40. Ibid., 13–14.

CHAPTER 4

1. *Report of the Commissioner of Indian Affairs Accompanying the Annual Report of the Secretary of the Interior, for the Year 1853* (Washington; A. O. P. Nicholson, 1854), 161. Hereafter cited as RCIA.

2. Ibid., 16.

3. Ibid., 121ff.

4. Ibid., 186.

5. Whitefield to Alfred Cumming, September 27, 1854, ibid., 90.

6. Rupert Richardson, *Comanche Barrier to the South Plains Settlement: A Century and a Half of Savage Resistance to the Advancing White Frontier* (Glendale, CA: Arthur H. Clark, 1933); Grant Foreman, *Advancing the Frontier, 1830–1860* (Norman: University of Oklahoma Press, 1933), 279; Ernest Wallace and E. Adamson Hoebel, *Comanche: Lords of the South Plains* (Norman: University of Oklahoma Press, 1952), 301; T. R. Fehrenbach, *Comanches: The Destruction of a People* (New York: Alfred A. Knopf, 1974), 388–91; David LaVere, *Contrary Neighbors: Southern Plains and Removed Indians in Indian Territory* (Norman: University of Oklahoma Press, 2000), 6, 126; Thomas W. Kavanagh, *Comanche: A History, 1706–1875* (Lincoln: University of Nebraska Press, 1996), 387; Andrew C. Isenberg, *Destruction of the Bison* (Cambridge, UK: Cambridge University Press, 2000) 110, 121–22; Elliott West, *Way to the West: Essays on the Central Plains* (Albuquerque: University of New Mexico Press, 1995), 62, 61, 65; Pekka Hämäläinen, *Comanche Empire* (New Haven, CT: Yale University Press, 2008), 299ff.

7. Tom D. Dillehay, "Late Quarternary Bison Population Changes on the Southern Plains," *Plains Anthropologist* 19 (1974): 185.

8. Timothy Baugh, "Ecology and Exchange: The Dynamics of Plains-Pueblo Interaction," in *Farmers, Hunters, and Colonists: Interaction Between the Southwest and the Southern Plains,* ed. Katherine Speilmann (Tucson: University of Arizona Press, 1991), 121.

9. Timothy Baugh, "Holocene Adaptations in the Southern High Plains," in *Plains Indians, A.D. 500–1500: The Archaeological Past of Historic Groups,* ed. Karl Schlesier (Norman: University of Oklahoma Press, 1994), 287.

10. John Speth, "Some Unexplored Aspects of Mutualistic Plains-Pueblo Food Exchange," in ibid., 27–28 for the discussion on Pueblo over hunting, and 29 for nutritional needs of a high protein diet.

11. Katherine Speilmann, "Coercion or Cooperation? Plains-Pueblo Interaction in the Protohistoric Period," in ibid., 40.

12. Fehrenbach, *Comanches,* 85.

13. Ibid., 87.

14. For an excellent discussion of the impact of the horse on Comanche life see ibid., 97–104.

15. Hämäläinen, *Comanche Empire,* 51ff.

16. Fehrenbach contends that the Sioux were not a factor in the Comanche migration southward but that the horses were, in *Comanches,* 130. Stanley Noyes, *Los Comanches: The Horse People, 1751–1845* (Albuquerque: University of New Mexico Press, 1993), xix; Kavanagh, *Comanche,* 63, 68. The most recent treatment of the Comanches by Pekka Hämäläinen, *Comanche Empire,* describes the Comanches as active agents in promulgating their own strategies and alliances in carving out their territory.

17. Pekka Hämäläinen, *Comanche Empire,* 40.

18. Ibid., 43.

19. Ibid., 49.

20. Ibid., 37.

21. Noyes, *Los Comanches,* xx.

22. James H. Shaw and Martin Lee, "Relative Abundance of Bison, Elk and Pronghorn on the Southern Plains, 1806–1857," *Plains Anthropologist* 42 (February 1997): 169.

23. West, *Way to the West,* 73–75.

24. *Message from the President of the United States Communicating the Discoveries made in Exploring the Missouri, Red River, and Washita by Captains Lewis and Clark, Doctor Sibley, and Mr. Dunbar* (Washington, DC: A. and G. Way, 1806), 75; Wallace and Hoebel, *The Comanche,* 74; Francis Levine, "Economic Perspectives on Comanchero Trade," in *Farmers, Hunters and Colonists,* ed. Katherine Speilmann, 157.

25. Andrew McKee Jones, "Comanche and Texans in the Making of the Comanche Nation: The Historical Anthropology of Comanche-Texan Relations, 1803–1997," (PhD dissertation, University of Wisconsin–Madison, 1997), 122–23.

26. Ibid., 122.

27. Flores, "Bison Ecology and Bison Diplomacy," 476.

28. Fehrenbach, *Comanches,* 388.

29. John C. Ewers, "The Influence of Epidemics on the Indian Populations and Cultures of Texas," in *Native American Demography in the Spanish Borderlands,* ed. Clark Spencer Larsen (New York: Garland Publishing Company, 1991), 170.

30. Senate, *Report of Captain R. B. Marcy,* S. Doc. 54, 32nd Cong., 2nd sess., 1853, 86–87. (Serial 666).

31. This discussion is contained in the first and second chapters of West, *Way to the West.*

32. Noah Smithwick, *Evolution of a State: Or Recollections of Old Texas Days* (Austin: Gammel Book Co., 1900), 188.

33. Kavanagh, *Comanche,* 345.

34. Richardson, *Comanche Barrier,* 173–74.

35. Latrobe, *Rambler in North America,* 203

36. Armstrong to Crawford, September 30, 1841, RCIA, 316.

37. George F. Ruxton, *Adventures in Mexico and the Rocky Mountains* (London: J. Murray, 1847), 266.

38. Ibid., 299.

39. Marcy quoted in Foreman, *Advancing the Frontier,* 261.

40. *Standard* (Clarksville, TX), May 28, 1853, in Richardson, *Comanche Barrier,* 175.

41. Ibid., 174.

42. Dan Flores, "Bison Ecology and Bison Diplomacy," 479.

43. West, *Way to the West,* 62; Dan Flores, *Horizontal Yellow: Nature and History in the Greater Southern Plains* (Albuquerque: University of New Mexico Press, 1999), 241.

44. West, *Way to the West,* 64.

45. Ibid., 39.

46. Wallace and Hoebel, *Comanche,* 296.

47. Ruxton, *Adventures in Mexico and the Rocky Mountains,* 266.

48. National Geophysical Data Center, at www.ngdc.noaa.gov/paleo/pdsiyear.html.

49. Joseph R. Smith, Diary of Joseph Smith, Western History Collections, University of Oklahoma, 10 and 23.

50. Ibid., 21 and 22.

51. Heth to Henry Stanton, August 4, 1851, Consolidated Correspondence File, Quartermaster General's Office, National Archives and Record Administration, Record Group 92, quoted in Leo E. Oliva, "Fort Atkinson on the Santa Fe Trail, 1850–1854," *Kansas Historical Quarterly* 40 (Summer 1974): 222.

52. Thomas Fitzpatrick to L. Lea, September 22, 1851, RCIA, 71.

53. Charles Halleck, "The Siege of Fort Atkinson," *Harper's New Monthly Magazine* 15 (October 1957): 638, quoted in Oliva, "Fort Atkinson on the Santa Fe Trail," 217.

54. Thomas Fitzpatrick to L. Lea, September 22, 1851, RCIA, 71.

55. Whitefield to Cumming, September 27, 1854, RCIA, 230.

56. Whitefield to Manypenny, September 4, 1855, RCIA, 117.

57. Drew to Manypenny, March 24, 1855, in Foreman, *Advancing the Frontier,* 282.

58. W. H. Garrett to C. W. Dean, August 24, 1855, RCIA, 135.

59. Richardson, *Comanche Barrier,* 212–13.

60. Whitefield to Cumming, September 27, 1854, RCIA, 91.

61. Mildred Mayhall, *Kiowas* (Norman: University of Oklahoma Press, 1987), 215.

62. Robert C. Mill to John Havery, October 114, 1857, RCIA, 143; J. L. Collins to C. E. Mix, September 27, 1858, RCIA, 185–86.

63. LaVere, *Contrary Neighbors,* 156.

64. Wallace and Hoebel, *Comanche,* 302; and F. Todd Smith, *Caddos, Wichitas, and the United States* (College Station: Texas A&M University Press, 1996), 52.

65. Lydia Caroline Baggett Jones (1843–1884), "Reminiscences," Western History Collections, University of Oklahoma.

66. John Chadburn Irwin, "Frontier Life of John Chadburn Irwin: From His Own Narrative as Given to Hazel Best Overton," 2, John Chadburn Irwin Papers, Southwest Collection, Texas Tech University, Lubbock, Texas.

67. *Waco (TX) Register,* April 21, 1866.

Chapter 5

1. T. N. Palmer to W. W. Lowe, January 10, 1859, Sen. Exec. Doc. 1, part 2, 36th Cong., 1st sess., 602.

2. Averam Bender, *The March of Empire: Frontier Defense in the Southwest, 1848–1860* (New York: Greenwood Publishers, 1968, c. 1952), 207.

3. Kenneth F. Neighbours, "Robert S. Neighbors and the Founding of Texas Indian Reservations" *West Texas Historical Association Yearbook* 31 (1955): 67–68.

4. Kenneth F. Neighbours, "Chapters from the History of Texas Indian Reservations" *West Texas Historical Association Yearbook* 33 (1957): 5.

5. Bender, *March of Empire,* 208.

6. Neighbours, "Robert S. Neighbors and the Founding of Texas Indian Reservations," 72.

7. Bender, *March of Empire,* 208–9; Rupert N. Richardson, "The Comanche Reservation in Texas," *West Texas Historical Association Yearbook* 5 (1929): 51–52.

8. Neighbours, "Chapters from the History of Texas Indian Reservations," 4–5.

9. Ibid., 8–9, 13.

10. George Klos, "'Our People Could not Distinguish One Tribe from Another': The 1859 Expulsion of the Reserve Indians from Texas," *West Texas Historical Association Yearbook* 97 (April 1994): 605.

11. Ibid., 10; Robert Neighbors to Charles E. Mix, September 10, 1855, in RCIA, 178.

12. Robert Neighbors to George Manypenny, September 18, 1856, RCIA, 68.

13. Robert Neighbors to George Manypenny, September 18, 1856, RCIA, 174.

14. Smith, *Caddos, Wichitas, and the United States,* 51.

15. Rupert N. Richardson, *The Frontier of Northwest Texas, 1846 to 1876* (Glendale, CA: Arthur H. Clark, 1963), 139.

16. Jerry D. Thompson, *Colonel Robert John Baylor: Texas Indian Fighter and Confederate Soldier* (Hillsboro, TX: Hill Junior College Press, 1971), 8; "To Control Blind Staggers in Horses," *Pacific Rural Press* 87, no. 11 (March 14, 1914), 341.

17. Ibid., 4.

18. Ibid., 7.

19. John Baylor to R. S. Neighbors, September 12, 1856, RCIA, 177.

20. Thompson, *Colonel John R. Baylor,* 11.

21. Raymond Estep, "Lieutenant William E. Burnet Letter: Removal of the Texas Indians and the Founding of Fort Cobb," *Chronicles of Oklahoma* 38, no. 4 (1960): 373.

22. Carl C. Rister, *Robert E. Lee in Texas* (Norman: University of Oklahoma Press, 1946), 85.

23. Kenneth F. Neighbours, *Robert Simpson Neighbors and the Texas Frontier* (Waco, TX: Texian Press, 1975), 214.

24. Smith, *Caddos, the Wichitas, and the United States,* 46.

25. Ibid., 55.

26. Ibid., 54.

27. S. P. Ross to Robert Neighbors, September 6, 1858, RCIA, 18.

28. Neighbours, "Chapters in the History of Texas Indian Reservations," 13.

29. Bender, *March of Empire,* 213–14.

30. Ibid., 213; Richardson, "The Comanche Reservation in Texas," 63.

31. Richardson, "The Comanche Reservation in Texas," 65.

32. Ibid., 65; Walter Prescott Webb, *The Texas Rangers: A Century of Frontier Defense* (Austin: University of Texas Press, 1964), 169.

33. Klos, "'Our People Could not Distinguish One Tribe from Another,'" 606–607.

34. Smith, *Caddos, the Wichitas, and the United States,* 57.

35. Richardson, "The Comanche Reservation in Texas," 66.

36. LaVere, *Contrary Neighbors,* 155–56.

37. S. P. Ross to Robert Neighbors, September 6, 1858, RCIA, 181.

38. S. P. Ross to Robert Neighbors, January 26, 1859, RCIA, 228.

39. *Dallas Herald,* September 8, 1858.

40. Robert Neighbors to George Manypenny, September 16, 1854, RCIA, 158–59.

41. J. J. Sturm to S. P. Ross, January 15, 1859, Sen. Exec. Doc. 1, part 2, 599.

42. J. J. Sturm to S. P. Ross, December 28, 1858, ibid., 589.

43. W. W. Cochran, B. F. Walker, J. Pollard, John Hitson, Jesse Hitson, Preston Witt, J. H. Dillahunty, and C. T. Hazlewood to W. W. Lowe, Major Neighbors, and Captain Ross, December 27, 1858, ibid., 608.

44. Proclamation by the Governor Executive Office Austin Texas, January 10, 1859, in *The Indian Papers of Texas and the Southwest,* vol. 3, ed. Dorman H. Winfrey and James M. Day (Austin: the Pemberton Press, 1966), 312.

45. Letter from H. Runnels to J. S. Ford, February 11, 1859, ibid., 314.

46. Robert Neighbors to J. W. Denver, January 15, 1859, Sen. Exec. Doc. 1, pt. 2, 590.

47. John S. Ford to E. J. Gurley, January 22, 1859, ibid., 605.

48. Robert Marlin to the People of Texas, January 4, 1859, ibid., 609.

49. Richardson, "The Comanche Reservation in Texas," 67.

50. Klos, "'Our People Could not Distinguish One Tribe from Another,'" 612.

51. "Letter to H. Runnels," May 24, 1859, *Indian Papers of Texas,* vol. 3, 328–29.

52. Raymond Estep, "Lieutenant William E. Burnet Letters: Removal of the Texas Indians and the Founding of Fort Cobb," *Chronicles of Oklahoma* 38 (1960), 370.

53. Ibid.

54. "Appointment of Peace Commission by H. R. Runnels," June 6, 1859, *Indian Papers of Texas,* vol. 3, 331–32.

55. John Henry Brown, G. B. Erath, J. M. Steiner, J. M. Smith, Richard Coke to H. R. Runnels, 1859, RCIA, 299.

56. J. H. Brown to R. S. Neighbors, July 14, 1859, *Indian Papers of Texas,* vol. 3, 334; R. S. Neighbors to J. H. Brown, July 17, 1859, *Indian Papers of Texas,* vol. 3, 335.

57. J. H. Brown to H. R. Runnels, July 22, 1859, *Indian Papers of Texas,* vol. 3, 339.

58. Kenneth F. Neighbours, "Indian Exodus out of Texas," *West Texas Historical Association Yearbook* 36 (October 1960): 82.

59. Ibid.

60. Estep, "Lieutenant William E. Burnet Letters," 376.

61. Ibid., 378.

62. Neighbours, "Indian Exodus out of Texas," 95–6.

63. Kenneth F. Neighbours, "The Assassination of Robert S. Neighbors," *West Texas Historical Association Yearbook* 34 (October 1958): 39.

64. Estep, "Lieutenant William E. Burnet Letters," 379.

65. Frances Mayhugh Holden, *Lambshead Before Interwoven: A Texas Range Chronicle, 1848–1878* (College Station: Texas A&M University Press, 1982), 86.

CHAPTER 6

1. *Waco Register,* April 21, 1866.

2. James Arnold, *Jeff Davis's Own: Cavalry, Comanches, and the Battle for the Texas Frontier* (New York: Wiley, 2000), 64; M. L. Crimmins, "Camp Cooper and Fort Griffin, Texas," in *Tracks Along the Clear Fork: Stories from Shackelford and Throckmorton Counties,* ed. Lawrence L. Cayton and John H. Farmer (Abilene, TX: McWhinney Foundation Press, 2000), 56; T. R. Fehrenbach, *Lone Star: A History of Texas and the Texans* (Boston: Da Capo, 1968; reprint, 2000), 523.

3. W. Jones to H. R. Runnels, October 17, 1858, *Indian Papers of Texas,* vol. 3, 297.

4. H. Ryan to H. R. Runnels, October 30, 1858, ibid., 301.

5. Gilbert Fite, *Farmer's Frontier, 1865–1900* (Norman: University of Oklahoma Press, 1987), 195.

6. Campbell, *Gone to Texas,* 266–67.

7. Arnold, *Jeff Davis's Own,* 121.

8. William C. Holden, "West Texas Drouths," *Southwestern Historical Quarterly* 32 (October 1928): 104. Holden cites the *Taylor County News,* June 6, 1886. See also Richard King, *Wagon's East: the Great Drouth of 1886,—an Episode in Natural Disaster, Human Relations, and Press Leadership* (Austin: the School of Journalism Development Program University of Texas, 1965), 3.

9. Joseph McConnell, *West Texas Frontier; or A Descriptive History of Early Times in Western Texas,* 2 vols. (Palo Pinto: Texas Legal Bank and Book Company, 1933), 285.

10. Arnold, *Jeff Davis's Own,* 78.

11. Washington Hammet to Captain Conner, February 19, 1860, ibid., 14.

12. John Williams to Sam Houston, February 20, 1860, ibid., 15.

13. J. H. Conner to Sam Houston, February 20, 1860, ibid., 16–17.

14. Charles L. Kenner, *A History of New Mexican–Plains Indian Relations* (Norman: University of Oklahoma Press, 1969), 129–32.

15. Walter Buenger, "Texas and the Riddle of Secession," in *Lone Star: Blue and Gray,* ed. Ralph A. Wooster (Austin: Texas State Historical Association, 1995), 5.

16. Ibid., 15.

17. Walter L. Buenger, *Secession and the Union in Texas* (Austin: University of Texas Press, 1984), 116.

18. A. W. Sparks, *The War Between the States as I saw it,* 9–10.

19. William W. White, "The Texas Slave Insurrection of 1860," *Southwestern Historical Quarterly* 52 (January 1949): 259–61.

20. Ibid., 262.

21. Buenger, "Texas and the Riddle of Secession," 12.

22. Ibid., 142–44, 148.

23. David Paul Smith, *Frontier Defense in the Civil War: Texas Rangers and Rebels* (College Station: Texas A&M University Press, 1992), 23–24.

24. Ibid., 24.

25. Jeanne T. Heidler, "'Embarrassing Situation': David E. Twiggs and the Surrender of United States Forces in Texas, 1861," in *Lone Star,* ed. Ralph A. Wooster, 39–40.

26. Carl Coke Rister, *Robert E. Lee in Texas* (Norman: University of Oklahoma Press, 1946), 160.

27. Smith, *Frontier Defense in the Civil War,* 25.

28. Heidler, "'Embarrassing Situation,'" 42–44.

29. Thomas T. Smith, *The U.S. Army and the Texas Frontier Economy, 1845–1900* (College Station: Texas A&M University Press, 1999), 53.

30. Ibid., 54.

31. All of these examples are discussed at length in Smith, *U.S. Army and the Texas Frontier Economy,* 132–34, 76–77, 89, and 128–31.

32. Ibid., 72, 92, 104, 110, and 114.

33. Ibid., 174–75.

34. Colonel J. F. K. Mansfield's Report of the Inspection of the Department of Texas in 1856, *Southwestern Historical Quarterly* 42 (March 1939): 372, 374, and 367.

35. Report from T. A. Washington, 1st Lt., 1st Infantry, Fort Chadbourne, to Major D. H. Vinton, San Antonio, May 28, 1857, records of the Fort Chadbourne Historic Site.

36. Arnold, *Jeff Davis's Own,* 62.

37. Ibid., 111.

38. Harold B. Simpson, *Frontier Forts of Texas* (Waco, TX: Texian Press, 1966), 14.

39. Arnold, *Jeff Davis's Own,* 262. Arnold cites Thomas to Assistant Adjutant General, December 15, 1859, microfilm 3348, Record Group 92, National Archives.

40. James W. C. Dechman to Sam Houston, May 16, 1860, *Indian Papers of Texas* vol 4, 34–35.

41. W. W. O. Stanfield to Sam Houston, ibid., 44.

42. Petition From the Citizens of Burnet County to Sam Houston, July 25, 1860, ibid., 38.

43. Martha Doty Freeman, Amy E. Dase, and Marie E. Blake, "Agriculture and Rural Development on Fort Hood Lands, 1849–1942: National Register Assessments of 710 Historical Archeological Properties" (U.S. Army Fort Hood: Environmental Management Office, 2001), 11.

CHAPTER 7

1. Arrell Gibson, *The Chickasaw* (Norman: University of Oklahoma Press, 1971), 103; and Muriel H. Wright, "Early Navigation and Commerce Along the Arkansas and Red Rivers of Oklahoma," *Chronicles of Oklahoma* 8 (March 1930): 77.

2. Ibid., 81.

3. Norman Arthur Graebner, "Pioneer Indian Agriculture in Oklahoma," *Chronicles of Oklahoma* 23 (Summer and Autumn 1945): 234.

4. Thomas Jefferson Farnham, *Travels in the Great Western Prairies, the Anahuac and Rocky Mountains, and in the Oregon Country,* in *Early Western Travels,* vol. 28, ed. Reuben Gold Thwaites (Cleveland, OH: Arthur H. Clark, 1906), 122.

5. Graebner, "Pioneer Indian Agriculture in Oklahoma," 235–36.

6. *Choctaw Intelligencer* (Doaksville, Choctaw Nation), October, 15, 1851, cited in Wright, "Early Navigation and Commerce," 84.

7. Monthly Meteorological Report, Fort Washita Meteorological Register, United States Weather Bureau Climatological Records, Record Group 27, Meteorological Records of the Surgeon General's Office 1819–1916, National Archives.

8. Douglas Cooper to Charles Dean, August 28, 1855, RCIA.

9. George W. Manypenny to R. McClelland, Secretary of the Interior, November 26, 1855, RCIA.

10. Michael F. Doran, "Antebellum Cattle Herding in the Indian Territory," *Geographical Review* 66 (January 1976): 54.

11. Foreman, *The Five Civilized Tribes,* 81.

12. Foreman, *Advancing the Frontier, 1830–1860,* 265.

13. George Manypenny to R. McClelland, November 26, 1855, RCIA, 9.

14. Douglas Cooper to Charles Dean, August 28, 1855, RCIA, 151.

15. J. C. Robinson to A. J. Smith, July 20, 1855, RCIA, 170.

16. John R. Whaley to his sister, November 1, 1855, "Letters of John R. Whaley," Archives and Manuscripts Division, Oklahoma Historical Society.

17. J. C. Robinson to D. H. Cooper, 1856, RCIA, 170.

18. Monthly Meteorological Report, Fort Washita Meteorological Register, National Archives, Record Group 27, Meteorological Records of the Surgeon General's Office.

19. George B. Hewitt, *Review of Factors Affecting Fecundity, Oviposition, and Egg Survival of Grasshoppers in North America,* [Beltsville, MD]: U.S. Department of Agriculture, Agricultural Research Service, 1985, 3–5.

20. Monthly Meteorological Report, Fort Washita Meteorological Register.

21. Foreman, *The Five Civilized Tribes,* 82.

22. C. W. Dean to George Manypenny, October 13, 1856, RCIA, 131–32.

23. Cyrus Byington to Douglas Cooper, August 3, 1857, ibid., 235.

24. J. W. Denver to J. Thompson, 1857, ibid., 6; W. H. Garrett to Elias Rector, September 21, 1857, ibid., 225; Ebenezer Hotchkin to Douglas Cooper, August 10, 1857, ibid., 245; J. C. Robinson to Douglas Cooper, August 21, 1857, ibid., 254.

25. Monthly Meteorological Report, Fort Washita Meteorological Register.

26. Elias Rector to Charles Mix, October 26, 1858, RCIA, 126.

27. "Famine in Kansas," *New York Times,* November 1, 1860.

28. George Butler to Elias Rector, September 10, 1859, RCIA, 178; W. H. Garrett to Elias Rector, September 12, 1859, ibid., 178; W. R. Baker to Douglas Cooper, July 27, 1859, ibid., 202.

29. Jesse McKee and Jon Schlenker, *The Choctaw: Cultural Evolution of a Native American Tribe* (Jackson: University Press of Mississippi, 1980), 142.

30. John Edwards, "The Choctaw Indians in the Middle of the Nineteenth Century," edited by John R. Swanton, *Chronicles of Oklahoma* 10 (1965): 408.

31. George Ainslie to Douglas Cooper, July 26, 1860, RCIA, 138.

32. Cyrus Kingsbury to Douglas Cooper, July 16, 1860, ibid., 137.

33. Anna Lewis, ed. "Diary of a Missionary to the Choctaw: 1860–1861," *Chronicles of Oklahoma* 17 (1939): 439–40.

34. Ibid., 440.

35. Elias Rector to A. B. Greenwood, September 24, 1860, RCIA, 117.

36. George W. Harkins to Peter Pitchlynn, February 10, 1860, box 3, folder 58, Pitchlynn Collection, Western History Collections, University of Oklahoma. Hereafter cited as Pitchlynn Collection.

37. Tandy Walker to Peter Pitchlynn, March 18, 1860 box 3, folder 63, ibid.

38. William Dole to Caleb Smith, November 27, 1861, box 3, folder 14, ibid.

39. Paul Bonnefield, "The Choctaw Nation on the Eve of the Civil War," *Journal of the West* 12 (July 1973): 388.

40. Joseph Dukes to Peter Pitchlynn, September 5, 1860, box 3, folder 81, Pitchlyn Collection.

41. L. P. Pitchlynn to Peter Pitchlynn, July 26, 1860, box 3, folder 77, ibid.; and Sampson Folsom to Peter Pitchlynn, October 8, 1860, box 3, folder 84, ibid.

42. A. B. Greenwood to J. Thompson, September 27, 1860, RCIA.

43. Graebner, "Pioneer Indian Agriculture in Oklahoma," 246.

44. Elias C. Boudinot to Stand Watie, February 12, 1861, in *Cherokee Cavaliers: Forty Years of Cherokee History as Told in the Correspondence of the Ridge-Watie-Boudinot Family,* ed. Edward E. Dale and Gaston Litton (Norman: University of Oklahoma Press, 1939), 103.

45. Ibid.

46. Angie Debo, *Rise and Fall of the Choctaw Republic,* 2nd ed. (Norman: University of Oklahoma Press, 1961), 74.

47. Annie H. Abel, *American Indian as Slaveholder and Secessionist* (Lincoln: University of Nebraska Press, 1992), 146.

48. Dean Trickett, "The Civil War in the Indian Territory 1861," *Chronicles of Oklahoma* 17 (September 1939): 315.

49. Paul Bonnefield, "Choctaw Nation on the Eve of the Civil War," 400.

50. Ibid.

51. McKee and Schlenker, *Choctaw,* 111.

52. Edwards, "Choctaw Indians in the Middle of the Nineteenth Century," 73 and 82; Debo, *The Rise and Fall of the Choctaw,* 81–82.

53. James Harrison to Edward Clark, April 23, 1861, in *War of the Rebellion: Official Records of the Union and Confederate Armies,* series 4, vol. 1, ed. Robert Scott, 322.

54. Ibid., 323.

55. Ibid., 323–24.

56. Ibid., 324–25.

57. Ibid., 323; Albert Pike to Robert Toombs, May 29, 1861, ibid., 359.

58. Ibid., 360.

59. Walter Lee Brown, *Life of Albert Pike* (Fayetteville: University of Arkansas Press, 1997), 359.

60. Ibid., 363.

61. Ibid., 364.

62. Articles of Convention, *War of the Rebellion,* series 4, vol. 1, 542–54.

63. Brown, *Life of Albert Pike,* 359.

64. Gibson, *Chickasaw,* 228.

65. Abel, *Indians as Slaveholders,* 58.

66. Larry C. Rampp and Donald L. Rampp, *Civil War in the Indian Territory* (Austin, TX: Presidial Press, 1975), 3.

67. Albert Pike to R. W. Johnson, May 11, 1861, *War of the Rebellion,* series 4, vol. 1, 457–58.

68. Albert Pike to R. W. Johnson, "Confederate Correspondence, etc.," May 11, 1861, *War of the Rebellion,* series 1, vol. 3, 572.

69. James D. Richardson, ed., *Messages and Papers of Jefferson Davis and the Confederacy Including Diplomatic Correspondence, 1861–1865* (New York: Chelsea House—Robert Hector Publishers, 1966), 150.

70. Fischer, *Civil War Era in Indian Territory,* 55; *War of the Rebellion,* series 4, vol. 1, 457 and 460; James F. Morgan, "The Choctaw Warrants of 1863," *Chronicles of Oklahoma* 57 (Spring 1979): 57.

71. "Secession of Indian Tribes," *New York Times,* July 14, 1861.

72. "Indians and the Government," *New York Times,* December 22, 1861.

73. Clara Sue Kidwell, *The Choctaw: A Critical Bibliography* (Bloomington: Indiana University Press, 1980), 46.

74. Cyrus Kingsbury to Reverend L. B. Treat, November 23, 1858, folder 5, file 50, Kingsbury Collection, Western History Collections, University of Oklahoma.

CHAPTER 8

1. Annie H. Abel, *The American Indian in the Civil War* (Lincoln: University of Nebraska Press, 1992), 44.

2. Ibid., 75, 45.

3. Report of Colonel Douglas H. Cooper, *War of the Rebellion,* series 1, vol. 8, 5.

4. Resolutions of the Senate and House of Representatives of the Chickasaw Legislature Assembled, May 25, 1861, ibid., vol. 3, 586.

5. Dean Trickett, "Civil War in the Indian Territory, 1861 (Continued)," *Chronicles of Oklahoma* 18 (June 1940): 150–52.

6. Pike to Benjamin, December 25, 1861, *War of the Rebellion,* series 1, vol. 8, 719–20.

7. Mary Jane Warde, "Now the Wolf Has Come: The Civilian Civil War in the Indian Territory," *Chronicles of Oklahoma* 71 (1993): 69.

8. Report of Colonel Douglass H. Cooper, January 20, 1862, *War of the Rebellion,* series 1, vol. 8, 5.

9. Ibid., 6.

10. Ibid.

11. Ibid., 7.

12. Ibid., 8.

13. Report of Colonel John Drew, December 18, 1861, ibid., 17.

14. Report of Colonel Douglas H. Cooper, ibid., 8–9.

15. Copper to McIntosh, December 11, 1861, ibid., 709.

16. Report of Colonel James McIntosh, January 1, 1861, ibid., 22.

17. Ibid., 24.

18. Report of Colonel Douglas H. Cooper," ibid., 13.

19. Dean Banks, "Civil War Refugees from Indian Territory in the North, 1861–1864," *Chronicles of Oklahoma* 41 (Autumn 1963): 289.

20. Warde, "Now the Wolf Has Come," 70.

21. Brown, *A Life of Albert Pike,* 379–80.

22. Ibid., 384.

23. Pike to Headquarters of Indian Territory, May 4, 1862, *War of the Rebellion,* series 1, vol. 13, 819–20.

24. Robert L. Duncan, *Reluctant General: the Life and Times of Albert Pike* (New York: E. P. Dutton and Co., 1961), 206–7.

25. Trickett, "Civil War in the Indian Territory," *Chronicles of Oklahoma* 19 (December 1941): 387–88.

26. Report of Brigadier General Albert Pike, March 14, 1862, *War of the Rebellion,* series 1, vol. 8, 288.

27. John Noble to Samuel Curtis, April 12, ibid., 206.

28. Duncan, *Reluctant General,* 225.

29. Wiley Britton, *Memoirs of the Rebellion on the Border—1863* (Chicago: Cushing, Thomas, 1882), 201–2.

30. Pike to J. S. Roane, June 1, 1862, *War of the Rebellion,* series 1, vol. 13, 935.

31. Pike to Secretary of War, June 26, 1862, ibid., 841–44.

32. Ibid., 843.

33. Pike to Randolph, June 27, 1862, ibid., 847.

34. Ibid., June 30, 1862, 848.

35. Pike to Hindman, June 8, 1862, ibid., vol. 13, 941–42.

36. Ibid., July 3, 1862, 957.

37. Reports of Major General T. C. Hindman, C. S. Army of Operations, May 31–November 3, 1862, ibid., 38.

38. Pike to Hindman, June 24, 1862, ibid., 947.

39. Pike to Secretary of War, July 20, 1862, ibid., 859–60.

40. Blunt to Weer, July 12, 1862, ibid., 488–89.

41. Pike to General S. Cooper, October 24, 1862, ibid., 893.

42. Pike to Hindman, July 3, 1862, ibid., 957–58; and Hindman to Curtis, November 3, 1862, ibid., 46.

43. Weer to Moonlight, July 12, ibid., 488.

44. Copper to Hindman, January 8, 1863, ibid., vol. 22, pt. 2, 770.

45. Pike to Anderson, October 26, 1862, ibid., 905.

46. Salomon to Blunt, July 20, ibid., 484–85.

47. Blunt to Schofield, November 9, ibid., 786.

48. Hannah Hicks, "Diary of Hannah Hicks," *American Scene* 13 (1972); see entry for November 17 and endnote 9.

49. Warde, "Now the Wolf Has Come," 73.

50. Ibid. Hannah Hicks, "Diary of Hannah Hicks," entries for December 18, 1862, and January 9, 1863.

CHAPTER 9

Epigraph. Ivan Tannehill, *Drought, Its Causes and Effects* (Princeton, NJ: Princeton University Press, 1947), vii.

1. Hindman to Pike, July 8, 1862, *War of the Rebellion,* series 1, vol. 13, 857.

2. To the Chiefs and the People of the Cherokee, Creek, Seminole, Chickasaw and Choctaw, ibid., 869–71.

3. Cooper to Hindman, August 7, 1862, ibid., 977.

4. Schofield to Phillips, January 11, 1863, ibid., vol. 22, 33.

5. Blunt to Phillips, February 23, 1863, ibid., 121.

6. Phillips to Curtis, January 19, 1863, ibid., 58.

7. Crosby to Cooper, January 27, 1863, ibid., 777.

8. Steele, to Anderson, January 27, 1863, ibid., 776.

9. Phillips to Curtis, January 29, 1863, ibid., 85.

10. Phillips to Curtis, February 4, 1863, ibid., 97.

11. Phillips to Curtis, February 15, 1863, ibid., 112.

12. Phillips to Blunt, March 3, ibid., 1863, 140.

13. Britton, *Memoirs of the Rebellion,* 175.

14. Phillips to Curtis, February 11, 1863, *War of the Rebellion,* series 1, vol 22, 109.

15. Steele to Cooper, January 18, 1863, ibid., 775.

16. Steele to Commanding Officer United States Forces, March 13, 1863, ibid., 154.

17. Steele to Anderson, March 15, 1863, ibid., 796.

18. Britton, *Memoirs of the Rebellion,* 198.

19. Seddon to Smith, March 18, 1863, *War of the Rebellion,* series 1, vol. 22, 802.

20. Steele to Lane, April 13, 1863, ibid., 827; Steele to Wigfall, April 15, 1863, ibid., 820.

21. Moty Kinnaird et al. to His Excellency Jefferson Davis, May 18, 1863, ibid., 1118ff.

22. Paul T. Wilson, "Delegates of the Five Tribes to the Confederate Congress," *Chronicles of Oklahoma* 53 (1975): 358.

23. Bassett to Gallaher, May 14, 1863, *War of the Rebellion,* vol. 53, 469.

24. "Report of Colonel William Phillips," 1863, ibid., vol. 22, 337.

25. Phillips to Blunt, May 15, 1863, ibid., 283.

26. Report of Brigadier-General Douglas H. Cooper, August 12, 1863, ibid., 458.

27. Report of Major-General James Blunt, July 26, 1863, ibid., 448.

28. Steele to Blair, July 22, 1863, ibid., 941.

29. Steele to Blair, August 3, 1863, ibid., 953.

30. Steele to Blair, July 22, 1863, ibid., 941.

31. Steele to Blair, August 7, 1863, ibid., 956.

32. Bankhead to Turner, August 20, 1863, ibid., 972–73.

33. Blunt to Curtis, December 30, ibid., 1862, part 1, 168.

34. Alfred Wade to Peter Pitchlynn, September 4, 1863, box 3, folder 113, Pitchlynn Collection.

35. Alfred Wade to Peter Pitchlynn, December 2, 1863, box 4, folder 7, ibid.

36. Alfred Wade to Peter Pitchlynn, February 7, 1863, box 3, folder 116, ibid.

37. Thomas Anderson to Sarah Watie, October 27, 1863, in *Cherokee Cavaliers*, 142.

38. Report of Brigadier General William Steele, February 15, 1864, *War of the Rebellion*, series I, vol. 22, 28.

39. Ibid., 34.

40. Ibid., 34.

41. Ibid., 35.

42. McNeil to Totten, January 4, 1864, ibid., vol. 34, 24.

43. Report of Colonel William A. Phillips, February 1, 1864, ibid., 107.

44. DuBose to Maxey, February 25, 1864, ibid., 998; Maxey to Smith, February 26, 1864, ibid., 996.

45. Ibid., 994.

46. Hardin to Adair, May 26, 1864, in *Cherokee Cavaliers*, 162.

47. Stand Watie to Sarah Watie, April 24, 1864, ibid., 156.

48. Report of Colonel Stand Watie, June 17, 1864, *War of the Rebellion*, vol. 41, 1012; and Watie to Headquarters First Indian Brigade, June 27, 1864, ibid., 1013.

49. S. B. Maxey Circular, June 15, 1864, ibid., vol. 34, 679.

50. Tandy Walker to Headquarters 2nd Indian Brigade, June 23, 1864, ibid., 694; and H. T. Martin to Headquarters of Indian Territory, June 27, ibid., vol. 41, 1013.

51. Maxey to Cooper, June 28, 1864, ibid., vol. 34, 697–98.

52. Portlock to Maxey, July 8, 1864, ibid., vol. 41, 998.

53. Maxey to Headquarters District of Indian Territory, July 30, 1864, ibid., 29.

54. Ibid., August 6, 1864, 29.

55. Cooper to Headquarters Department of Indian Territory, August 10, 1864, ibid., 36.

56. Boggs to Price, August 4, 1864, ibid., 729.

57. Maxey to Headquarters District of Indian Territory, October 1, 1864, ibid., 779.

58. Report of Colonel Colton Greene, December 18, 1864, ibid., 692.

59. Pitchlynn to Maxey, December 29, 1864, ibid., vol. 53, 1035.

60. Ibid.

61. Maxey to Pitchlynn, January 10, 1865, box 4, folder 18, Pitchlynn Collection.

62. T. M. Scott to Stand Watie, March 22, 1865, in *Cherokee Cavaliers*, 220.

63. Copper to Pitchlynn, April 25, 1865, box 4, folder 18, Pitchlynn Collection.

64. Coleman to Bussey, March 31, 1865, *War of the Rebellion*, series I, vol. 48, 16.

65. Cooper to Scott, May 17, 1865, ibid., 1311.

66. Stand Watie to Sarah Watie, May 27, 1865, in *Cherokee Cavaliers*, 227.

67. Veatch to Crosby, July 20, 1865, *War of the Rebellion*, series I, vol. 48, 1096.

68. Cooper to Buckner, June 28, 1865, ibid., 1098.

69. Blunt to Levering, June 7,1865, ibid., 805–6.

70. Blunt to Curtis, April 22, 1863, ibid., vol. 22, 235.

71. MaKean to Chapman, April 11, 1864, ibid., vol. 34, 137.

72. Annual Report of Milo Gookins for 1864, H.R. Exec. Doc. 1, 39th Cong., 1st sess., 319–20.

73. General Order No. 27, September 6, 1864, *War of the Rebellion*, series 1, vol. 41, 86.

74. Cutler to Blunt, May 27, 1865, ibid., vol. 48, 636–37.

75. Gookins to Superintendent of Indian Affairs, May 16, 1865, H.R. Exec. Doc. 1, 39th Cong., 1st sess., 450, Serial set 1248, fiche 6.

76. Ibid., 437, serial set 1248, fiche 5.

77. Report of C. C. Snow, September 25, 1865, ibid., 477, fiche 6.

78. Jerald C. Walker, "The Difficulty of Celebrating an Invasion," in *"An Oklahoma I Had Never Seen Before": Alternative Views of Oklahoma History*, ed. Davis D. Joyce (Norman: University of Oklahoma Press, 1980), 18.

79. Kingsbury to Treat, December 4, 1865, box 5, folder 4, Kingsbury Collection, Western History Collections, University of Oklahoma.

Chapter 10

Epigraph. Max Krueger, *Pioneer Life in Texas* (San Antonio, Press of the Clegg Company, 1930), 184.

1. "More Mail Robberies," *Dallas Herald*, September 1, 1858.

2. "Horrible Encounter," ibid., September 8, 1858.

3. John Baylor to Fran, April 28, 1857, John R. Baylor Papers, Eugene C. Barker Library, University of Texas.

4. Andrew C. Isenberg, *The Destruction of the Bison* (New York City, Cambridge University Press, 2000), 100.

5. Ibid., 100–103.

6. Jo Ella Powell Exley, *Texas Tears and Sunshine: Voices of Frontier Women*, ed. (College Station: Texas A&M University Press, 1985), 79. Also see Gary Anderson, *Conquest of Texas: Ethnic Cleansing in the Promised Land, 1820–1875* (Norman: University of Oklahoma Press, 2005), 138.

7. Letter from Lampasas Citizens to Senators Burney and Cooley, August 15, 1866, *Indian Papers of Texas*, vol. 4, 103–4.

8. J. H. Chrisman to Sam Houston, February 11, 1860, ibid., 3.

9. Alexander Walters to Sam Houston, Feb. 12, 1860, ibid., 5–6.

10. Oliver Loving to F. R. Lubbock, 1862, ibid., 67.

11. Report of Lorenzo Labadi, August 23, 1867, in RCIA, 1867, 214.

12. J. Y. Dashiel to F. R. Lubbock, August 12, 1863, James Buckner Barry Papers, Center for American History Collection at the Barker Library, University of Texas. Hereafter cited as Barry Papers.

13. General Smith to Cooper, May 28, 1851, 2855, Letters Received, Adjutant General's Office, Record Group 94, National Archives.

14. Philip Deloria, *Playing Indian* (New Haven, CT: Yale University Press, 1998), 11.

15. Anderson, *Conquest of Texas*, 288.

16. *Dallas Herald,* July 24, 1858.

17. Ibid., September 1, 1858.

18. Ibid., August 21, 1858.

19. Ibid., September 8, 1858.

20. Cashion, *Texas Frontier,* 48, cites *Dallas Herald,* September 1, 1858.

21. S. A. Blain to Sam Houston, April 23, 1860, in *Indian Papers of Texas,* vol. 4, 32.

22. Ibid., 56–63.

23. Ibid., 56–63

24. McCulloch to Barry, September 12, 1861, Barry Papers.

25. David Pickering and Judy Falls, *Brush Men and Vigilantes: Civil War Dissent in Texas* (College Station: Texas A&M University Press, 2000), 68.

26. Roberts to Turner, August 29, 1863, *War of the Rebellion,* series 1, vol. 26, 187.

27. Magruder to Murrah, December 21, 1863, ibid., 519.

28. Smith, *Frontier Defense in the Civil War,* 88, 99, and 102.

29. Ibid., 77.

30. Report Written to Governor Edward Clark of Texas, April 23, 1861, *War of the Rebellion,* series 4, vol. 1, 323.

31. McCulloch to Magruder, October 21, 1863, *War of the Rebellion,* series 1, vol. 26, 345.

32. General Order Number Seven, January 12, 1864, *War of the Rebellion,* series 1, vol. 34, 856.

33. Smith, *Frontier Defense,* 123–24.

34. Ibid., 124.

35. Henry McCulloch to E. P. Turner, March 15, 1864, *War of the Rebellion,* series 1, vol. 34, 1045.

36. Henry Fossett to Colonel J. B. Barry, June 1, 1864, Barry Papers.

37. Smith, *Frontier Defense in the Civil War,* 151.

38. Throckmorton to Murrah, December 9, 1864, Governor Pendleton Murrah Records, Texas State Library.

39. William Quayle to Governor Murrah, December 27, 1863, Adjutant General's Papers, General Correspondence, Texas State Library.

40. Petition of Citizens of Gillespie, Kerr and Kimble Counties, March 31, 1864, box 401-386, folder 386-15, Adjutant General's Records, Texas State Library.

41. H. T. Edgar to A. G. Dickenson, May 11, 1864, *War of the Rebellion,* series 1, vol. 34, 817.

42. Lockhart to Dickenson, ibid., 818.

43. McAdoo to Culberson, September 1864, box 410-387, folder 387-4, Adjutant General's Records, Texas State Library.

44. George A. Forsyth to general P. H. Sheridan, November 2, 1866, in S. Doc. 19, 45th Cong., 2nd sess., 1: 8–9, serial 1780.

Chapter 11

Epigraph. John Opie, *Ogallala: Water for a Dry Land* (Lincoln: University of Nebraska Press, 1993), 68.

1. James C. Malin, "Dust Storms: Part Two, 1861–1880," *Kansas Historical Quarterly* 14 (August 1946): 265.

2. Walter Prescott Webb, *Great Plains,* 391.

3. David Emmons, *Garden in the Grasslands,* 15, and 17–18.

4. Ibid., 18.

5. Henry Nash Smith, "Rain Follows the Plow: The Notion of Increased Rainfall for the Great Plains, 1844–1880," *Huntington Library Quarterly* 10, no. 2 (February 1947): 177.

6. W. H. Droze, "Changing the Plains Environment: The Afforestation of the Trans-Mississippi West," *Agricultural History* 51 (January 1977): 9.

7. Henry Nash Smith, "Rain Follows the Plow," 180.

8. Ibid.; and Emmons, *Garden in the Grasslands,* 143, 145–46.

9. Thomas R. Wessel, "Prologue to the Shelterbelt, 1870 to 1934," *Journal of the West* 6 (January 1967): 125.

10. Ibid.

11. Nathanial H. Egelston, "Report of the Chief of Forestry," *United States Department of Agriculture Report, 1883,* 453; in Wessel, "Prologue to the Shelterbelt," 124.

12. Droze, "Changing the Plains Environment," 16.

13. Wessel, "Prologue to the Shelterbelt," 126.

14. Ibid., 51.

15. Ibid., 60, 62.

16. Ibid., 25.

17. Ibid., 25–35.

18. Eleanor L. Turk, "Selling the Heartland: Agents, Agencies, Press, and Policies Promoting German Emigration to Kansas in the Nineteenth Century," *Kansas History* 12 (Autumn 1989): 158.

19. Rupert N. Richardson, *The Greater Southwest: The Economic, Social, and Cultural Development of Kansas, Oklahoma, Texas, Utah, Colorado, Nevada, New Mexico, Arizona, and California from the Spanish Conquest to the Twentieth Century* (Glendale, CA: Arthur H. Clark, 1934), 360.

20. Jan Blodgett, *Land of Bright Promise: Advertising the Texas Panhandle and South Plains* (Austin: University of Texas Press, 1988), 27.

21. Ibid., 26.

22. Rathjen, *Texas Panhandle Frontier,* 245.

23. Blodgett, *Land of Bright Promise,* 28–9.

24. Rupert N. Richardson, *Greater Southwest,* 349–50.

25. Jonathon Richards to Enoch Hoag, September 1, 1874, RCIA, 238; Jonathon Richards to Edward Smith, September 1, 1875, ibid., 288.

26. John Shorb to Commissioner of Indian Affairs, September 4, 1880, ibid., 90–91. See also John Smith to Commissioner of Indian Affairs, October 25, 1879, ibid., 71.

27. William Whiting to Commissioner of Indian Affairs, August 31, 1880, ibid., 84.

28. Jonathon Richards to E. P. Smith, September 1, 1875, ibid., 288; William Nicholson to Commissioner of Indian Affairs, September 22, 1876, ibid., 71; A. C. Williams to Commissioner of Indian Affairs, August 20, 1877, ibid., 112; and A. C. Williams to Commissioner of Indian Affairs, August 31, 1878, ibid., 70.

29. Levi Woodward to Commissioner of Indian Affairs, August 27, 1877, ibid., 105; and A. C. Williams to Commissioner of Indian Affairs, August 24, 1877, ibid., 68.

30. Carl Coke Rister, "Free Land Hunters of the Southern Plains," *Chronicles of Oklahoma* 22 (Winter 1944–1945): 399.

31. Carl Coke Rister, "Oklahoma, the Land of Promise," *Chronicles of Oklahoma* 23 (Spring 1945): 4.

32. Ibid.

33. H. Craig Miner, "Cherokee Sovereignty in the Gilded Age: The Outlet Question," *Chronicles of Oklahoma* 71 (Summer 1993): 128.

34. Rister, "Oklahoma, Land of Promise," 5.

35. Ibid., 6.

36. Stan Hoig, *Fort Reno and the Indian Territory Frontier* (Fayetteville: University of Arkansas Press, 2000), 85.

37. Rister, "Oklahoma, Land of Promise," 13.

38. James C. Malin, "Dust Storms, Part III, 1880–1900" *Kansas Historical Quarterly* 14 (November 1946): 398.

39. J. W. Williams, "A Statistical Study of the Drouth of 1886," *West Texas Historical Association Yearbook* 21 (October 1945): 92.

40. Ibid., 97.

41. Holden, "West Texas Drouths," 105.

42. Ibid., 106.

43. Ibid., 107.

44. Holden, "West Texas Drouths," 106.

45. Arrell Morgan Gibson, "Ranching on the Southern Great Plains," *Journal of the West* 6 (January 1967): 149.

46. Ibid., 150.

47. Andy Addington, Indian Pioneer History Conservation Commission (IPHCC), Grant Foreman Papers, vol. 1, Oklahoma Historical Society, 61. Mr. Addington uses the years 1884 and 1885 for his account, but was obviously confusing these years with the well-documented drought and winter of 1886–1887.

48. Williams, "A Statistical Study of the Drouth of 1886," 93.

49. Holden, "West Texas Drouths," 111–12.

50. Ibid., 110–11.

51. Ibid., 111.

52. Williams, "A Statistical Study of the Drought of 1886," 100.

53. Ibid., 102; and Holden, "West Texas Drouths," 119.

54. Ibid., 112.

55. Ibid., 112–13.

56. Ibid., 120–21; and Williams, "A Statistical Study of the Drought of 1886," 103–104.

57. Malin, "Dust Storms," 400.

58. Ibid., 401.

59. Elizabeth Brooks and Jacque Emel, "The Llano Estacado of the American Southern High Plains," in *Regions At Risk: Comparisons of Threatened Environments,* eds. Jeanne Kasperson, Roger Kasperson, and B. L. Turner (New York: United Nations University Press, 1995), 264.

60. Fite, *The Farmer's Frontier, 1865–1900,* 200.

61. *Albany (TX) News,* October 21, 1886; in Williams, "A Statistical Study of the Drought of 1886," 105.

62. Jesse Ausubel and Asit K. Biswas, *Climatic Constraints and Human Activities* (New York: Pergamon, 1980), 109.

63. Ibid., 105.

64. Holden, "West Texas Drouths," 122.

65. Atchison, Topeka and Santa Fe Railroad Company, *How and Where to Get a Living. A Sketch of "the Garden of the West."* (Boston: Atchison, Topeka and Santa Fe Railroad Company, 1876), 36.

66. Union Pacific, *The Resources and Attraction of Nebraska* (Omaha: Union Pacific Railroad, 1893), 8; in Emmons, *Garden in the Grasslands,* 149.

67. Ibid., 153.

68. Malin, "Dust Storms," 403.

69. Robert S. Winslow, Diaries, July 11, September 11, and September 30, 1893, Southwest Collection, Texas Tech University, 3.

70. "The Passing of Louis Hill," *Albany (TX) News,* April 1, 1932.

71. Alvin O. Turner, "Order and Disorder: The Opening of the Cherokee Outlet," *Chronicles of Oklahoma* 71 (Summer 1993): 160.

72. Author's emphasis. Webb and Hill to Henry C. Payne, August 3, 1893, Webb and Hill Letter Press Book, Louis Hamilton Hill Papers, Southwest Collection, Texas Tech University, 377.

73. Webb and Hill to L. S. Kohrnhorst, August 4, 1893, ibid., 387.

74. Webb and Hill to James Thompson, September 20, 1893, ibid., 126.

75. Webb and Hill to A. J. Vick, September 20, 1893, ibid., 124.

76. Ibid., and Webb and Hill to Col. Alvin Rockwell, September 21, 1893, ibid., 138.

77. Webb and Hill to Evans, Snider, Buel Company National Stockyards, Chicago, IL, October 3, 1893, ibid., 295–96.

78. *Albany (TX) News,* April 1, 1932.

79. Charles Alling Interview, IPHCC, vol. 12, 296–97.

80. Howard L. Johnson and Claude E. Duchon, *Atlas of Oklahoma Climate* (Norman: University of Oklahoma Press, 1995), fig. 3.1.

81. Irene Richy Interview, IPHCC, vol. 15, 41.

82. Frank T. Perry Interview, IPHCC, vol. 53, 299.

83. USDA, Weather Bureau, 1894, reel. no. 1.176, 24–36.

84. Malin, "Dust Storms," 404.

85. Ibid., 405.

86. Ibid., 409.

87. Malin, "Dust Storms," 409.

88. Fite, *Farmer's Frontier,* 212.

89. Bonnie Lynn-Sherow, "Ordering the Elements: An Environmental History of West Central Oklahoma," (PhD dissertation, Northwestern University, Evanston, IL, 1998): 115.

90. Richard S. Cutter, Diary, March 17, 1894, Richard S. Cutter Papers, Southwest Collection, Texas Tech University.

91. Ibid., July 13, 1894.

92. R. Warrick and M. Bowden, "The Changing Impacts of Drought in the Great Plains," in *The Great Plains: Perspectives and Prospects,* ed. M. Lawson and M. Baker (Lincoln: University of Nebraska Press, 1980), 126.

93. *Hennessey (Okla. Terr.) Clipper,* March 19, 1896.

94. *Hardesty (Okla. Terr.) Herald,* June 15, 1896.

95. *El Reno (Okla. Terr.) News,* September 11, 1896.

96. A. E. Woodson to the Commissioner of Indian Affairs, September 17, 1894, RCIA, 236.

97. H. B. Freeman to the Commissioner of Indian Affairs, August 18, 1894, ibid., 242; and J. P. Woolsey to the Commissioner of Indian Affairs, August 15, 1894, ibid., 253.

98. Ibid., 247.

99. A. H. Viets to the Superintendent of Indian Schools, July 1, 1895, ibid., 247.

100. W. J. A. Montgomery to the Superintendent of Indian Schools, July 1, 1895, ibid., 248.

101. Cora M. Dunn to the Commissioner of Indian Affairs, August 25, 1895, ibid., 253.

102. J. P. Brown to the Superintendent of Indian Schools, June 30, 1895, ibid., 263.

103. A. H. Viets to the Commissioner of Indian Affairs, June 30, 1896, ibid., 252.

104. Ibid.

105. A. E. Woodson to the Commissioner of Indians Affairs, August 28, 1896, ibid., 249.

106. Oklahoma Agricultural Experiment Station, "Irrigation for Oklahoma" Bulletin No. 18 (April 1896): 3–11.

107. A. H. Viets to the Commissioner of Indian Affairs, June 30, 1896, ibid., 252; and Franklin Baldwin to the Commissioner of Indian Affairs, August 28, 1896, ibid., 253.

108. Sherow, "Ordering the Elements," 118.

109. Department of the Interior, "Report on the Population of the United States at the Twelfth Census: 1900" (Washington, DC: GPO, 1905).

Chapter 12

Epigraph. Dwight Morrow, from an October 1930 speech quoted in *Encarta Book of Quotations,* ed. Bill Swainson (New York: St. Martin's Press, 2000), 672.

1. Donna A. Barnes, *Farmers in Rebellion: the Rise and Fall of the Southern Farmers' Alliance and People's Party in Texas"* (Austin: University of Texas Press, 1984), 1.

2. J. Frank Norris, *The Octopus: A Story of California* (New York: Doubleday, Page, 1903).

3. Lawrence Goodwin, *Populist Moment: A Short History of the Agrarian Revolt in America* (New York: Oxford University Press, 1978),16–17.

4. Ibid., 26.

5. Mark W. Harrington, "Weather Making, Ancient and Modern, *National Geographic* 6 (August 1894): 44.

6. Emmons, *Garden in the Grasslands,* 135.

7. *Watonga (Okla. Terr.) Republican,* May 15, 1895.

8. Ibid., 148.

9. Plutarch, *Plutarch's Lives,* vol. 9, *Demetrius and Antony, Pyrrhus and Caius Marius* (Cambridge, MA: Harvard University Press, 1920; reprint, 1959), 521.

10. Everett Dick, *Conquering the Great American Desert: Nebraska* (Lincoln: Nebraska State Historical Society, 1975), 339.

11. Jeff A. Townsend, "Nineteenth and Twentieth Century Rainmaking in the United States," (master's thesis, Texas Tech University, 1975), 26.

12. Walter Prescott Webb, *Great Plains,* 380n2.

13. Harrington, "Weather Making, Ancient and Modern," 55.

14. Ibid., 56.

15. Ibid., 380–81.

16. Walter Prescott Webb, "Some Vagaries of the Search for Water in the Great Plains," *Panhandle-Plains Historical Review* 3 (1930): 35.

17. Clark C. Spence, *Rainmakers: American Pluviculture to World War II* (Lincoln: University of Nebraska Press, 1980), 55–61.

18. Martha B. Caldwell, "Some Kansas Rainmakers," *Kansas Historical Quarterly* 7 (August 1938): 309–11.

19. Ibid., 312.

20. Harrington, "Weather Making, Ancient and Modern," 48.

21. Spence, *Rainmakers,* 70–71.

22. Harrington, "Weather Making, Ancient and Modern," 319.

23. *Beaver Advocate* (Okla. Terr.), July 27, 1893.

24. A. Bower Sageser, "Editor Bristow and the Great Plains Irrigation Revival of the 1890's," *Journal of the West* 3 (January 1964): 83.

25. Webb, *Great Plains,* 382.

26. "Miscellaneous News," *Herald* (Iowa City, IA), September 1, 1894.

27. Krueger, *Pioneer Life in Texas,* 190.

28. Ibid., 184.

29. Ibid., 191.

30. Ibid., 192.

31. Ibid.

32. Ibid., 195.

33. Mark Wahlgren Summers, *The Gilded Age: or, The Hazard of New Functions* (Upper Saddle River, NJ: Prentice Hall, 1997), 239–40.

34. Henry M. Littlefield, "The Wizard of Oz: Parable on Populism," *American Quarterly* 16, no. 1 (Spring 1964): 47–58.

CHAPTER 13

Epigraph. Clyde Muchmore, "I Saw the Run," in *The Last Run, Kay County, Oklahoma, 1893,* ed. Daughters of the American Revolution (Ponca City, OK: Courier Printing, 1934), 120.

1. L. G. Moses, *Wild West Show and the Images of American Indians, 1883–1933* (Albuquerque, University of New Mexico Press, 1996), 130.

2. William Cronin, *Nature's Metropolis: Chicago and the Great West* (New York: W. W. Norton and Co., 1991), 342.

3. Henry Nash Smith, *Virgin Land: The American West as Symbol and Myth* (Cambridge, MA: Harvard University Press, 1978), 201–2; Frederick Jackson Turner, *The Significance of the Frontier in American History* (New York City: Penguin Books, 2008), 99ff.

4. Berlin Chapman, "Opening of the Cherokee Outlet: An Archival Study," *Chronicles of Oklahoma* 40 (Summer and Autumn 1962): 159. Historians have led the way in introducing the negative aspects of the Cherokee Outlet Land Run. Joe Milam has argued that special interest groups, most notably the railroads, were able to gain the opening of the outlet at the expense of the Cherokees, Pawnees, and Tonkowas. He examined the hardships endured by those who made the run and praised the toughness and determination of those who were able to stick it out. Joe B. Milam, "The Opening of the Cherokee Strip," *Chronicles of Oklahoma* 9 (September and December 1931): 454–75.

In 1962, Berlin Chapman documented the paper trail that led to the opening and the attempts by the federal government to prevent fraudulent access to claims. Berlin B.

Chapman, "The Opening of the Cherokee Outlet: An Archival Study," *Chronicles of Oklahoma* 40 (Summer and Autumn 1962): 158–81, 253–85. A whole volume of the *Chronicles of Oklahoma* has been devoted to the Cherokee Outlet, including articles highlighting the challenges the event posed to Cherokee Nation sovereignty, the role of cattle corporations in the struggle to open or keep closed the area to settlement, and a discussion on the pattern of lawlessness and disorder that marked the run. Alvin O. Turner, "Order and Disorder," *Chronicles of Oklahoma* 71, no. 2 (1993): 154–73.

Popular interpretations of the land runs have tended to portray the opening of the Cherokee Outlet as a time of opportunity for settlers. A host of communities in the Outlet re-enact the land run and celebrate the egalitarian aspect of granting land to the fastest settlers to arrive. Earl Newsom has published a book which has the most in common with the public perception of the event. His work is celebratory in spirit and does a fine job of researching the background to the formation of the outlet. After discussing the day of the run, he goes on to detail the history of several of the larger towns of the outlet with booster-like exuberance. D. Earl Newsom, *Cherokee Strip: Its History and Grand Opening* (Stillwater, OK: New Forms Press, 1992), 1–172.

5. "Strip Boomer," *Norman Transcript,* March 24, 1893.

6. Newsom, *Cherokee Strip,* 18.

7. William W. Savage, Jr., *Cherokee Strip Livestock Association: Federal Regulation and the Cattleman's Last Frontier* (Columbia: University of Missouri Press, 1973), 47ff.

8. Craig Miner, "Cherokee Sovereignty in the Gilded Age: The Outlet Question," *Chronicles of Oklahoma* 71, no. 2 (Summer 1993): 128.

9. Ibid., 130.

10. William W. Savage, Jr., "Of Cattle and Corporations: The Rise, Progress, and Termination of the Cherokee Livestock Association," *Chronicles of Oklahoma* 71, no. 2 (Summer 1993): 147.

11. "All Ready," *Daily Oklahoma State Capital* (Guthrie), September 11, 1893.

12. Chapman, "The Opening of the Cherokee Outlet," 168.

13. Turner, "Order and Disorder," 166.

14. Newsom, *Cherokee Strip,* 36.

15. Chapman, "Opening of the Cherokee Outlet," 173; Mary Holton Boerner, ed., *Run of '93* (Ponca City, OK: The '93er Association), 156.

16. Turner, "Order and Disorder," 163.

17. Ibid., 160.

18. *Guthrie Daily Leader,* September 12, 1893.

19. Ibid.; "Boomers Broil," *Daily Oklahoma State Capital* (Guthrie), September 15, 1893.

20. Chapman, "Opening of the Cherokee Outlet," 257.

21. Boerner, *Run of '93,* 165.

22. "Water for Strippers," *Guthrie Daily Leader,* September 14, 1893.

23. Boerner, *Run of '93,* 132.

24. *Guthrie Daily Leader,* September 16, 1893.

25. These sites included Kirk, Kildare, Cross, and Wharton (Perry). Ibid., September 14, 1893.

26. "U.S. Department of Agriculture Weather Bureau," *Beaver Advocate,* September 14, 1893.

27. "A Request," *Daily Oklahoma State Capital,* September 14, 1893.

28. "How to Catch Sooners," *Guthrie Daily Leader,* September 14, 1893.

29. Ibid.

30. "Orlando Registration," *Daily Oklahoma State Capital,* September 15, 1893.

31. Ibid.

32. Milam, "Opening of the Cherokee Outlet," 469.

33. Lucille Gilstrap, "Homesteading the Strip," *Chronicles of Oklahoma* 51 (Fall 1973): 288.

34. Boerner, *Run of '93,* 194.

35. Mary Bobbit Brown Hatfield, box H-45, folder 6, Edna Green Parker Collection, Western History Collections, University of Oklahoma, 1955, 1.

36. Interview with Etta Stocking, Indian-Pioneer History Collection, 103: 227, Archive Manuscript Division, Oklahoma Historical Society. Hereafter cited as IPH.

37. Boerner, *Run of '93,* 268.

38. J. S. Wade, "Uncle Sam's Horse Race," *Chronicles of Oklahoma* 35 (Summer 1957): 149; interview with Andrew T. Anthony, IPH, 12: 383.

39. The numbers of registrants at the booths were 30,000 at Arkansas City; 15,000 at Caldwell; 15,000 at Orlando; 10,000 at Kiowa; 10,000 at Hunnewell; 10,000 at Stillwater; 10,000 at Hennessey; 5,000 at Goodwin; and 5,000 at Cameron, for a total of 110,000 registrants.

40. Chapman, "Opening of the Cherokee Outlet," 259; Turner, "Order and Disorder," 164.

41. *Guthrie Daily Leader,* September 21, 1893.

42. *Daily Oklahoma State Capitol,* September 15, 1893.

43. *Guthrie Daily Leader,* September 24, 1893.

44. "U.S. Department of Agriculture for Week Ending September 18, 1893," *Beaver Advocate,* September 28, 1893.

45. Ibid., September 22, 1893.

46. Boerner, *Run of '93,* 10.

47. *El Reno News,* July 24, 1896; *Beaver Advocate,* August 23, 1894; *Mulhall (Okla. Terr.) Chief,* September 21, 1894; ibid., August 3, 1894; and *Beaver Advocate,* August 16, 1894.

48. *Mulhall Chief,* August 3, 1894.

49. *Guthrie Daily Leader,* September 20, 1893.

50. Boerner, *Run of '93,* 232.

51. *Guthrie Daily Leader,* September 21, 1893.

52. Turner, "Order and Disorder," 170.

53. Kenny L. Brown, "Building a Life: Culture, Society, and Leisure in the Cherokee Outlet," *Chronicles of Oklahoma* 71 (Summer 1993): 180.

54. *Mulhall Chief,* August 3, 1894.

55. *Eagle Gazette* (Stillwater, Okla. Terr.), September 6, 1894.

56. *Hennessey Clipper,* April 23, 1896.

57. Ibid., November 26, 1896.

58. Boerner, *Run of '93,* 134.

59. Ibid., 143–47.

60. Ibid., 141.

61. "Governor's Message to the Third Legislative Assembly of the Territory of Oklahoma," January 8, 1895, Oklahoma Territory Governor's Messages and Reports, Archives Division, Oklahoma Department of Libraries, 9.

62. Interview with Frank T. Perry, IPH, 53: 299.

63. Interview with Mrs. R. D. Neal, IPH, 7: 433.

64. *Edmond Sun Democrat,* June 7, 1895.

65. Interview with Andrew Anthony, IPH, 12: 385–86.

66. Brown, "Building a Life," 192.

67. Kevin Sweeney, "And the Skies Were Not Cloudy All Day," *Chronicles of Oklahoma* 81 (Winter 2003–2004): 450–52.

68. Ibid.

69. Ibid.

70. Ibid.

71. Boerner, *Run of '93,* 223, 229.

72. Atchison, Topeka and Santa Fe Railroad Company, *Cherokee Strip and Oklahoma: Opening of Cherokee Strip, Kickapoo, Pawnee and Tonkawa Reservations* (Chicago: Poole Bros., 1893), 2–3.

73. Margaret A. Nelson, *Home On the Range* (Boston: Chapman and Grimes, 1947), 162–67.

EPILOGUE

Epigraph. Max Krueger, *Pioneer Life in Texas,* 184.

1. Ibid., 193–94.

2. James Earl Sherow, *Watering the Valley: Development Along the High Plains Arkansas River, 1870–1950* (Lawrence: University Press of Kansas, 1990), 13.

3. Ibid., 17.

4. Sageser, "Editor Bristow and the Great Plains Irrigation Revival," 78.

5. John Opie, *Ogallala: Water for a Dry Land* (Lincoln: University of Nebraska Press, 2000), 69.

6. Sageser, "Editor Bristow and the Great Plains Irrigation Revival," 85.

7. Donald Green, *Land of the Underground Rain: Irrigation on the Texas High Plains, 1910–1970* (Austin: University of Texas Press, 1973), 233.

8. Ibid., 234.

9. Opie, *Ogallala,* 63.

10. Green, *Land of the Underground Rain,* 232.

11. Donald J. Pisani, *To Reclaim a Divided West: Water Law, and Public Policy, 1848–1902* (Albuquerque: University of New Mexico Press, 1992), 273ff.

12. Donald Worster, *Dust Bowl: The Southern Plains in the 1930s* (New York: Oxford University Press, 1979), 89–94.

13. Ibid., 13–14.

14. Opie, *Ogallala,* 93.

15. Warrick and Bowden, "Changing Impacts of Droughts in the Great Plains," 127.

16. David M. Kennedy, *Freedom from Fear: The American People in Depression and War, 1929–1945* (New York: Oxford University Press, 1999), 153ff.

17. Opie, *Ogallala,* 353.

18. Tai Kreidler, "To Anchor the Wind," *Journal of the West* 29, vol. 4 (1990): 46–52.

19. Worster, *Dust Bowl,* 222.

20. Ibid., 223.

21. Johnson and Duchon, *Atlas of Oklahoma Climate,* 13; and Arthur H. Doerr, "Dry Conditions in Oklahoma in the 1930s and 1950s as Delimited by the Original Thornwaite Climatic Classification," *Great Plains Journal* 2 (Spring 1963): 75.

22. Opie, *Ogallala,* 107.

23. Ibid., 108.

24. Ibid., 126–28, 131.

25. Department of the Interior, *Report on the Population of the United States at the Eleventh Census: 1890* (Washington, DC: GPO, 1895), 12, 19–20, 31, 35, and 40–42.

26. Sherow, *Watering the Valley,* 167.

27. Dennis Dimick, "If You Think the Water Crisis Can't Get Worse, Wait Until the Aquifers are Drained," *National Geographic* (August 12, 2014), http://news.national geographic.com/news/2014/08/140819-groundwater-california-drought-aquifers-hidden -crisis/.

28. "U.S.–Brazil Cotton Deal Perpetuates an Unhealthy Status Quo of Subsidies," *Washington Post,* October 7, 2014, washingtonpost.com/opinions/us-brazil-cotton-deal -perpetuates-an-unhealthy-status-quo-of-subsidies/2014/10/07/d8346bf4-4b2a-11e4 -891d-713f052086a0_story.html.

Bibliography

DOCUMENTS

Baker, James H. Diary, 1856–1920. Barker History Center. University of Texas.

Cook, E. R., D. M. Meko and C. W. Stockton, eds. *U.S. Drought Area Index Reconstructions.* Boulder, CO: National Oceanic and Atmospheric Administration, National Geophysical Data Center Paleoclimatology Program, 1998. www.ngdc .noaa.gov/paleo/pdsiyear.html.

Cutter, Richard S. "Diary." Richard S. Cutter Papers. Southwest Collection. Texas Tech University.

Indian Pioneer History Conservation Commission. Grant Foreman Papers. Archive and Manuscript Department. Oklahoma Historical Society.

Gales and Seaton's Register of Debates in Congress, 1824–1837. 13 vols. Washington, DC: Government Printing Office, 1825–1839.

"Governor's Message to the Third Legislative Assembly of the Territory of Oklahoma." January 8, 1895. Oklahoma Territory Governor's Messages and Reports. Archives Division. Oklahoma Department of Libraries.

Hatfield, Mary Bobbit Brown. Edna Green Parker Collection. Box H-45 F6. Western History Collections. University of Oklahoma.

Hayden, Ferdinand Vandeveer. *First, Second, and Third Annual Reports of the United States Geological Survey of the Territories for the Years 1867, 1868, and 1869, Under the Department of the Interior.* Washington, DC: Government Printing Office, 1873.

Jefferson, Thomas. Message from the President of the United States Communicating the Discoveries made in Exploring the Missouri, Red River, and Washita by Captain Lewis and Clark, Doctor Sibley, and Mr. Dunbar. Washington, DC: A. and G. Way, 1806.

Jones, Lydia Caroline Baggett (1843–1884). "Reminiscences." Western History Collections. University of Oklahoma.

Oklahoma Agricultural Experiment Station. "Irrigation for Oklahoma." Bulletin No. 18 (April 1896): 3–11.

Report of the Commissioner of Indian Affairs Accompanying the Annual Report of the Secretary of the Interior, For the Year 1853. Washington, DC: A. O. P. Nicholson, 1841, 1854–1865.

Scott, Robert N. *The War of the Rebellion: A Compilation of the Official Records of the Union and Confederate Armies.* Harrisburg, PA: National Historical Society, 1985.

Smith, Joseph R. Diary of Joseph R. Smith. Western History Collections. University of Oklahoma.

U.S. Congress. House. *Country for the Indians West of the Mississippi.* H.R. Doc. 172, 22nd Cong., 1st sess., 1832.

———. *Preservation and Civilization of the Indians.* H.R. Doc. 231, 19th Cong., 1st sess., 1826.

———. *Removal Indians Westward.* H.R. Doc. 87, 20th Cong., 2nd sess., 1828.

U.S. Congress. Senate. *Captain Randolph B. Marcy, Report.* Exec. Doc. 64, 31st Cong., 1st sess., 1850.

———. *Exchange of Lands with the Indians.* S. Doc. 145, 14th Cong., 2nd sess., 1817.

———. Exec. Doc. 1, H 1, 35th Cong., 2nd sess., 1856.

———. *Journal of Lieutenant J. W. Abert, From Bent's Fort to St. Louis, in 1845.* S. Doc. 438. 29th Cong., 1st sess., 1846.

———. *Memorial of the Prudential Committee for the American Board of Commissioners for Foreign Missions.* S. Doc. 50, 21st Cong., 2nd sess., 1831.

———. *A Plan for Removing the Several Indian Tribes West of the Mississippi River.* S. Doc. 218, 18th Cong., 2nd sess., 1825.

———. *Report of Captain R. B. Marcy,* S. Doc. 54, 32nd Cong., 2nd sess., 1854.

———. *Report of the Committee on Indian Affairs.* S. Doc. 246, 24th Cong., 1st sess., 1836.

———. *Views of the Cherokees in Relation to Further Cessions of their Lands.* S. Doc. 208, 18th Cong., 1st sess., 1824.

U.S. Department of the Interior. *Report on the Population of the United States at the Twelfth Census: 1900.* Washington, DC: Government Printing Office, 1905.

U.S. Department of War. Records of the Adjutant General's Office, 1865–1875. Record Group 94. National Archives and Records Administration, Washington, DC. (Doc. ID 250-M-1851/9.)

U.S. Weather Bureau. Monthly Meteorological Reports. Climatological Records. Oklahoma Climatological Survey.

Whaley, John R. Letters of John R. Whaley. Archives and Manuscripts Division. Oklahoma Historical Society.

Webb and Hill Letter Press Book. Louis Hamilton Hill Papers. Southwest Collection. Texas Tech University.

Wheeler, Otis. "Letter to His Sister, 25 May 1855." Southwest Collection. Texas Tech University.

Winslow, Robert S. Diaries, Southwest Collection. Texas Tech University.

NEWSPAPERS

Albany (TX) News.
Beaver Advocate (Okla. Terr.).
Choctaw Intelligencer (Doaksville, Choctaw Nation).
Daily Oklahoma State Capital (Guthrie, Okla. Terr.).
Dallas (TX) Herald.
Eagle Gazette, (Stillwater, Okla. Terr.).
Edmond (Okla. Terr.) Sun Democrat.
El Reno (Okla. Terr.) News.

Guthrie (Okla. Terr.) Daily Leader.
Hardesty (Okla. Terr.) Herald.
Hennessey (Okla. Terr.) Clipper.
Iowa City (IA) Herald.
Kingfisher Free Press (Okla. Terr.).
Mulhall (Okla. Terr.) Chief.
Norman (Okla. Terr.) Transcript.
Sequoyah Memorial (Tahlequah, Cherokee Nation).
Topeka (KS) Daily Capital.
Watonga (Okla. Terr.) Republican.

BOOKS

Abel, Annie H. *The History of Events Resulting in Indian Consolidation West of the Mississippi.* Washington, DC: American Historical Association, 1906.
————. *The American Indian as Participant in the Civil War.* St. Clair Shores, MI: Scholarly Press, 1972.
————. *The American Indian as Slaveholder and Secessionist.* Lincoln: University of Nebraska Press, 1992.
————. *The American Indian and the End of the Confederacy.* Lincoln: University of Nebraska Press, 1993.
Arnold, James R. *Jeff Davis's Own: Cavalry, Comanches and the Battle for the Texas Frontier.* New York: Wiley, 2000.
Atchison, Topeka and Santa Fe Railroad Company. *Cherokee Strip and Oklahoma: Opening of the Cherokee Strip, Kickapoo, Pawnee and Tonkawa Reservations.* Chicago: Poole Brothers, 1893.
————. *How and Where to Get a Living: A Sketch of "the Garden of the West."* Boston: Santa Fe Railroad Company, 1876.
Ausubel, Jesse, and Asit K. Biswas. *Climatic Constraints and Human Activities.* New York: Pergamon Press, 1980.
Bell, John R. *The Journal of Captain John R. Bell.* Edited by LeRoy R. Hafen and Harlin M. Fuller. Glendale, CA: Authur H. Clark, 1957.
Bender, Averam B. *The March of Empire: Frontier Defense in the Southwest, 1846–1860.* New York: Greenwood Press, 1968, c. 1952.
Benton, Thomas Hart. *Discourse of Mr. Benton of Missouri before the Mercantile Library Association . . . Delivered in Tremont Temple at Boston, December 20, 1854.* Washington, DC: J. T. and L. Towers, 1854.
Billington, Ray Allen. *Westward Expansion: A History of the American Frontier.* New York: McMillan Co., 1949. Reprint, 1954.
Blodgett, Jan. *Land of Bright Promise: Advertising the Texas Panhandle and South Plains.* Austin: University of Texas Press, 1988.
Blanford, H. F. "How Rain is Formed." *Annual Report of the Board of Regents of the Smithsonian . . . 1889.* Washington, DC: Smithsonian Institute, 1890.
Boerner, Mary Holton, ed. *Run of '93.* Ponca City, OK: The 93'er Association, 1991.
Brackenridge, Henry M. *Journal of a Voyage up the Missouri River in Eighteen Hundred and Eleven.* In Reuben Gold Thwaites, ed. *Early Western Travels, 1748–1846,* vol. 6. Cleveland, OH: Arthur H. Clark, 1904.

———. *Views of Louisiana; Together with a Journal of a Voyage Up the Missouri River in 1811*. Pittsburgh, PA: Cramer, Spear and Richbaum, 1814.

Britton, Wiley. *Memoirs of the Rebellion on the Border—1863*. Chicago: Cushing, Thomas, 1882.

Buenger, Walter. *Secession and the Union in Texas*. Austin: University of Texas Press, 1984.

Campbell, Randolph B. *Gone to Texas: A History of the Lone Star State*. New York: Oxford University Press, 2003.

Cashion, Ty. *A Texas Frontier: The Clear Fork Country and Fort Griffin, 1849–1887*. Norman: University of Oklahoma Press, 1996.

Catlin, George. *Letters and Notes on the Manners, Customs, and Conditions of the North American Indians: Written During Eight Years' Travel Amongst the Wildest Tribes of Indians in North America*. Philadelphia, PA: J. W. Bradley, 1860.

Clary, David A. *Eagles and Empire: The United States, Mexico, and the Struggle for a Continent*. New York: Bantam Books, 2009.

Crayton, Lawrence L., and John H. Farmer, eds. *Tracks Along the Clear Fork: Stories from Shackelford and Throckmorton Countie*. Abilene, TX: McWhinney Foundation Press, 2000.

Cronon, William. *Nature's Metropolis: Chicago and the Great West*. New York: W. W. Norton, 1991.

Cunfer, Geoff. *On the Great Plains: Agriculture and Environment*. College Station: Texas A&M University Press, 2005.

Dale, Edward Everett, and Gaston Litton, eds. *Cherokee Cavaliers: Forty Years of Cherokee History as Told in the Correspondence of the Ridge-Watie-Boudinot Family*. Norman: University of Oklahoma Press, 1939. Reprint, 1940.

Daughters of the American Revolution, eds. *The Last Run: Kay County, Oklahoma, 1893*. Ponca City, OK: Courier Printing, 1934.

Debo, Angie. *The Rise and Fall of the Choctaw Republic*. Norman: Unversity of Oklahoma Press, 1934. Reprint, 1972.

———. *The Rise and Fall of the Choctaw Republic*, 2nd ed. Norman: University of Oklahoma Press, 1961.

De Rosier, Arthur H. *The Removal of the Choctaw Indians*. Knoxville: University of Tennessee Press, 1970.

Dick, Everett. *Conquering the Great American Desert: Nebraska*. Lincoln: Nebraska State Historical Society, 1975.

Duncan, Robert Lipscomb. *Reluctant General: The Life and Times of Albert Pike*. New York: E. P. Dutton, 1961.

Eisenhower, John S. D. *So Far From God: The U.S. War with Mexico, 1846–1848*. New York: Random House, 1989.

Ella, George M. *Isaac McCoy: Apostle of the Western Trail*. Springfield, MO: Particular Baptist Press, 2002.

Ellsworth, Henry L. *Washington Irving on the Prairies; or A Narrative of a Tour of the Southwest in the year 1832*. New York: American Book Company, 1937.

Emmons, David. *Garden in the Grasslands: Boomer Literature of the Central Plains*. Lincoln: University of Nebraska Press, 1971.

Evans, Howard Ensign. *The Natural History of the Long Expedition to the Rocky Mountains, 1819–1820*. New York: Oxford University Press, 1997.

Ewen, David, ed. *American Popular Songs: From the Revolutionary War to the Present*. New York: Random House, 1966.

Farnham, Thomas Jefferson. "Travels in the Great Western Prairies, the Anahuac and Rocky Mountains, and in the Oregon Country." In Reuben Gold Thwaites, ed. *Early Western Travels,* vol. 28. Cleveland, OH: Arthur H. Clark, 1906.

Fehrenbach, T. R. *Comanches: The Destruction of a People.* New York: Alfred A. Knopf, 1974.

———. *Lone Star: A History of Texas and the Texans.* Boston: Da Capo, 1968. Reprint, 2000.

Fisher, LeRoy H. *The Civil War in Indian Territory.* Los Angeles: Lorrin L. Morrison, 1974.

Fite, Gilbert. *The Farmer's Frontier, 1865–1900.* Norman: University of Oklahoma Press, 1987.

Flores, Dan. *Horizontal Yellow: Nature and History in the Near Southwest.* Albuquerque: University of New Mexico Press, 1999.

Freeman, Martha Doty. *Chronological History of Fort Chadbourne.* Dallas: Fort Chadbourne Foundation, 2002.

Freeman, Martha Doty, Amy E. Dase, and Marie E. Blake. *Agriculture and Rural Development on Fort Hood Lands, 1849–1942: National Register Assessments of 710 Historical Archeological Properties.* U.S. Army Fort Hood: Environmental Management Office, 2001.

Folsom, Joseph P., comp. and ed. *Constitution and Laws of the Choctaw Nation together with the Treaties of 1855, 1865, and 1866.* New York: William P. Lyon and Sons, 1869. Reprint 1973.

Foreman, Grant. *Advancing the Frontier, 1830–1860.* Norman: University of Oklahoma Press, 1933.

———. *The Five Civilized Tribes.* Norman: University of Oklahoma Press, 1934.

———. *Indian Removal: The Emigration of the Five Civilized Tribes of Indians.* Norman: University of Oklahoma Press, 1953.

Frazer, Robert Walter. *Forts and Supplies: The Role of the Army in the Economy of the Southwest, 1846–1861.* Albuquerque: University of New Mexico Press, 1983.

Gibson, Arrell Morgan. *The Chickasaws.* Norman: University of Oklahoma Press, 1971.

———. *Oklahoma: A History of Five Centuries,* 2nd ed. Norman: University of Oklahoma Press, 1981.

Gilpin, William. *Mission of the North American People: Geographical, Social, and Political,* 2nd ed. Philadelphia: J. B. Lippincott., 1874.

Goetzmann, William H. *Army Exploration in the American West, 1803–1863.* New Haven, CT: Yale University Press, 1959.

Goodman, George and Cheryl Lawson. *Retracing Major Stephen H. Long's 1820 Expedition: The Itinerary and Botany.* Norman: University of Oklahoma Press, 1995.

Green, Donald. *Land of the Underground Rain: Irrigation on the Texas High Plains, 1910–1970.* Austin: University of Texas Press, 1973.

Gregg, Josiah. *Commerce of the Prairies.* Ed. Max L. Moorhead. Norman: University of Oklahoma Press, 1954.

———. *Commerce of the Prairies: The Journal of a Santa Fe Trader.* In Reuben Gold Thwaites, ed. *Early Western Travels,* vol. 28. Cleveland, OH: Arthur H. Clark, 1906.

Haines, Francis. *The Plains Indian.* New York: Thomas Y. Crowell, 1976.

Hämäläinen, Pekka. *The Comanche Empire.* New Haven, CT: Yale University Press, 2008.

Hareven, T. K., ed. *Anonymous Americans: Exploration in the Nineteenth Century Social History.* Englewood Cliffs, NJ: Prentice Hall, 1971.

Hickerson, Nancy P. *The Jumanos: Hunters and Traders of the South Plains.* Austin: University of Texas Press, 1994.

Hoig, Stan. *Fort Reno and the Indian Territory Frontier.* Fayetteville: University of Arkansas Press, 2000.

Holden, Frances Mayhugh. *Lambshead Before Interwoven: A Texas Range Chronicle, 1848–1878.* College Station: Texas A&M University Press, 1982.

Hollon, W. Eugene. *The Great American Desert, Then and Now.* New York: Oxford University Press, 1966.

Holmes, William F., ed. *American Populism.* Lexington, MA: D. C. Heath, 1994.

Isenberg, Andrew C. *The Destruction of the Bison.* New York: Cambridge University Press, 2000.

James, Edwin. *From Pittsburgh to the Rocky Mountains: Major Stephen Long's Expedition, 1819–1820.* Edited by Maxine Benson. Golden, CO: Fulcrum, 1988.

———. *Part 1 of James's Account of S. H. Long's Expedition, 1819–1820.* In Reuben Gold Thwaites, ed. *Early Western Travels, 1748–1846,* vol. 14. Cleveland, OH: Arthur H. Clark, 1905.

———. *Part 2 of James's Account of S. H. Long's Expedition, 1819–1820.* In Reuben Gold Thwaites, ed. *Early Western Travels, 1748–1846,* vol. 15. Cleveland, OH: Arthur H. Clark, 1905.

———. *Part 3 of James's Account of S. H. Long's Expedition, 1819–1820.* In Reuben Gold Thwaites, ed. *Early Western Travels, 1748–1846,* vol. 16. Cleveland, OH: Arthur H. Clark, 1905.

———. *Part 4 of James's Account of S. H. Long's Expedition, 1819–1820.* In Reuben Gold Thwaites, ed. *Early Western Travels, 1748–1846,* vol. 17. Cleveland, OH: Arthur H. Clark, 1905.

Johnson, Howard L. and Claude E. Duchon. *Atlas of Oklahoma Climate.* Norman: University of Oklahoma Press, 1995.

Josephy, Alvin M., Jr. *The American Heritage Book of the Pioneer Spirit.* New York: American Heritage, 1959.

Kasperson, Jeanne, Roger Kasperson and B. L. Turner, eds. *Regions At Risk: Comparisons of Threatened Environments.* New York: United Nations University Press, 1995.

Kavanagh, Thomas W. *The Comanches: A History, 1706–1875.* Lincoln: University of Nebraska Press, 1996.

Kennedy, David M. *Freedom from Fear: The American People in Depression and War, 1929–1946.* New York: Oxford University Press, 1999.

Kenner, Charles L. *A History of New Mexican-Plains Indian Relations.* Norman: University of Oklahoma Press, 1969.

Kidwell, Clara Sue. *The Choctaws in Oklahoma: From Tribe to Nation, 1855–1970.* Norman: University of Oklahoma Press, 2007.

King, C. R. *Wagon's East.* Austin: University of Texas Press, 1965.

Krueger, Max. *Pioneer Life in Texas.* San Antonio: Clegg Company, 1930.

Larsen, Clark Spencer, ed. *Native American Demography in the Spanish Borderlands.* New York: Garland Publishing, 1991.

Latrobe, Charles. *The Rambler in North America, 1832–1833.* New York: Johnson Reprint Corp., 1970.

LaVere, David. *Contrary Neighbors: Southern Plains and Removed Indians in Indian Territory.* Norman: University of Oklahoma Press, 2000.

Lawson, Merlin P. *The Climate of the Great American Desert: Reconstruction of the Climate of Western Interior United States, 1800–1850.* Lincoln: University of Nebraska Press, 1974.

Lawson, Merlin P., and M. Baker, eds. *The Great Plains: Perspectives and Prospects.* Lincoln: University of Nebraska Press, 1980.

Lewis, Meriwether, and William Clark. *The Original Journals of the Lewis and Clark Expedition, 1804–1806.* Edited by Reuben Gold Thwaites. 8 vols. New York: Antiquarian Press, 1959.

Malin, James. *The Grasslands of North America: Prolegomena to its History.* Gloucester, MA: P. Smith, 1967.

———. *Indian Policy and Westward Expansion.* Lawrence: University of Kansas Press, 1921.

Malone, Dumas, ed. *Dictionary of American Biography.* New York: Scribner's Sons, 1946.

Martin, Geoffrey, and Preston James, eds. *All Possible Worlds: A History of Geographical Ideas.* New York: John Wiley and Sons, 1993.

Matthews, Sallie Reynolds. *Interwoven: A Pioneer Chronicle.* College Station: Texas A&M University Press, 1982.

Mayhall, Mildred P. *The Kiowas.* Norman: University of Oklahoma Press, 1987.

McCaslin, Richard B. *Tainted Breeze: The Great Hanging at Gainesville, Texas, 1862.* Baton Rouge: Louisiana State University Press, 1994.

McCoy, Isaac. *History of Baptist Indian Missions Embracing Remarks on the Former and Present Condition of the Aboriginal Tribes: Their Former Settlement Within the Indians Territory, and Their Future Prospects.* Washington, DC: William M. Morrison, 1840.

———. "Report on the Country for Indians West of the Mississippi River." In David A. White, ed. *News of the Plains and Rockies, 1803–1865.* Vol. 3, Spokane, WA: Authur H. Clark, 1997.

———. *Remarks on the Practicability of Indian Reform, Embracing Their Colonization.* New York: Gray and Bunce, 1829.

McDermott, John Francis. *The Western Journals of Washington Irving.* Norman: University of Oklahoma Press, 1966.

Mollhausen, Baldwin. *Diary of a Journey from the Mississippi to the Coasts of the Pacific.* New York: Johnson Reprint Corp., 1969.

Moring, John. *Men with Sand: Great Explorers of the North American West.* Helena, MT: Falcon Publishing, 1998.

Morse, Jedidiah. *A New Universal Gazeteer, or Geographical Dictionary Accompanied with an Atlas.* New Haven, CT: S. Converse, 1823.

———. *Report to the Secretary of War of the United States on Indian Affairs, Comprising a Narrative of a Tour Performed in the Summer of 1820, under a Commission from the President of the United States, for the Purpose of Ascertaining, for the Use of the Government, the Actual State of the Indian Tribes in Our Country.* New Haven, CT: S. Converse, 1822.

Moses, L. G. *The Indian Man: A Biography of James Mooney.* Urbana: University of Illinois Press, 1984.

Muchmore, Clyde. "I Saw the Run." In Daughters of the American Revolution, eds. *The Last Run: Kay County, Oklahoma, 1893.* Ponca City, OK: Courier Printing, 1934.

Neighbours, Kenneth. *Robert Simpson Neighbors and the Texas Frontier.* Waco, TX: Texian Press, 1975.

Nelson, Margaret A. *Home on the Range.* Boston: Chapman and Grimes, 1947.

Newsom, D. Earl. *The Cherokee Strip: Its History and Grand Opening.* Stillwater, OK: New Forms Press, 1992.

Nichols, James L. *The Confederate Quartermaster in the Trans-Mississippi.* Austin: University of Texas Press, 1964.

Nichols, Roger L., and Patrick L. Halley. *Stephen Long and American Frontier Exploration.* Norman: University of Oklahoma Press, 1995.

Norris, J. Frank. *The Octopus: A Story of California.* New York: Doubleday, Page, 1903.

Noyes, Stanley. *Los Comanches: The Horse People, 1751–1845.* Albuquerque: University of New Mexico Press, 1993.

Opie, John. *Ogallala: Water for a Dry Land,* 2nd ed. Lincoln: University of Nebraska Press, 2000.

Payne, William Morton. *Leading American Essayists.* New York: Books for Libraries, 1910. Reprint, 1968.

Pike, Zebulon Montgomery. *Pike's Dissertation of Louisiana.* In Donald Jackson, ed. *The Journals of Zebulon Montgomery Pike,* vol. 2. Norman: University of Oklahoma Press, 1966.

Pisani, Donald J. *To Reclaim a Divided West: Water, Law, and Public Policy, 1848–1902.* Albuquerque: University of New Mexico Press, 1992.

Plutarch. *Plutarch's Lives.* Vol. 9, *Demetrius and Antony, Pyrrhus and Caius Marius.* Cambridge, MA: Harvard University Press, 1920. Reprint, 1959.

Powell, John Wesley. *Report on the Lands of the Arid Region of the United States.* Edited by Wallace Stegner. Cambridge, MA: Harvard University Press, 1962.

Pownall, Thomas. *A Topographical Description of the Dominions of the United States of America: [Being a Revised and Enlarged Edition of] A Topographical Description of Such Parts of North America as are Contained in the (Annexed) Map of the Middle British Colonies, &c. In North America.* Pittsburgh, PA: University of Pittsburgh Press, 1949.

Prucha, Francis Paul, ed. *Cherokee Removal: The "William Penn" Essays and Other Writings.* Knoxville: University of Tennessee Press, 1981.

Rampp, Larry C., and Donald L. Rampp. *The Civil War in the Indian Territory.* Austin, TX: Presidential Press, 1975.

Rathjen, Frederick W. *The Texas Panhandle Frontier.* Lubbock: Texas Tech University Press, 1973.

Richardson, James D., ed. *The Messages and Papers of Jefferson Davis and the Confederacy Including Diplomatic Correspondence, 1861–65.* Vol. 1. New York: Chelsea House—Robert Hector Publishers, 1966.

Richardson, Rupert Norval. *The Comanche Barrier to the South Plains Settlement: A Century and a Half of Savage Resistance to the Advancing White Frontier.* Glendale, CA: Arthur H. Clark, 1933.

———. *The Greater Southwest: The Economic, Social, and Cultural Development of Kansas, Oklahoma, Texas, Utah, Colorado, Nevada, New Mexico, Arizona, and California from the Spanish Conquest to the Twentieth Century.* Glendale, CA: Arthur H. Clark, 1934.

———. *The Frontier of Northwest Texas, 1846 to 1876.* Glendale, CA: Arthur H. Clark, 1963.

Ruxton, George F. *Adventures in Mexico and the Rocky Mountains.* London: J. Murray, 1847.

Savage, William W., Jr. *The Cherokee Strip Livestock Association: Federal Regulation and the Cattleman's Last Frontier.* Columbia: University of Missouri Press, 1973.

Say, Thomas. *Account of an Expedition from Pittsburgh to the Rocky Mountains.* Edited by Edwin James. Philadelphia, PA: Carey and Lea, 1823.

Schleiser, Karl, ed. *Plains Indians, A.D. 500–1500: The Archaeological Past of Historic Groups.* Norman: University of Oklahoma Press, 1994.

Schultz, George A. *An Indian Canaan: Isaac McCoy and the Vision of an Indian State.* Norman: University of Oklahoma Press, 1972.

Sherow, James Earl. *Watering the Valley: Development Along the High Plains Arkansas River, 1870–1950.* Lawrence: University Press of Kansas, 1990.

Simmons, Marc, ed. *Border Comanches: Seven Spanish Colonial Documents, 1785–1819.* Santa Fe: Stagecoach Press, 1967.

Simpson, Harold B. *Frontier Forts of Texas.* Waco: Texian Press, 1966.

Smith, David Paul. *Frontier Defense in the Civil War: Texas Rangers and Rebels.* College Station: Texas A&M University Press, 1992.

Smith, F. Todd. *The Caddos, the Wichitas, and the United States, 1846–1901.* College Station: Texas A&M University Press, 1996.

Smith, Thomas T. *The U.S. Army and the Texas Frontier Economy, 1845–1900.* College Station: Texas A&M University Press, 1999.

Smithwick, Noah. *The Evolution of a State: Or Recollections of Old Texas Days.* Austin: Gammel Book Co., 1900.

Sparks, A. W. *The War Between the States, As I Saw it: Reminiscent, Historical, Personal.* Longview, TX: D and D Publishing, 1987.

Speilmann, Katherine. *Farmers, Hunters, and Colonists: Interaction Between the Southwest and the Southern Plains.* Tucson: University of Arizona Press, 1991.

Spence, Clark C. *The Rainmakers: American Pluviculture to World War II.* Lincoln: University of Nebraska Press, 1980.

Thompson, Jerry D. *Colonel John Robert Baylor: Texas Indian Fighter and Confederate Soldier.* Hillsboro, TX: Hill Junior College Press, 1971.

Thornton, Russell. *The Cherokee: A Population History.* Lincoln: University of Nebraska Press, 1990.

Thwaites, Reuben Gold, ed. *Early Western Travels, 1748–1846,* vols. 6, 14–17. Cleveland, OH: Arthur H. Clark, 1904.

Turner, Frederick Jackson. *The Significance of the Frontier in American History.* New York City, Penguin Books: 2008.

Wallace, Ernest, and E. Adamson Hoebel. *The Comanches: Lords of the South Plains.* Norman: University of Oklahoma Press, 1952.

Webb, Walter Prescott. *The Great Plains.* Dallas: Ginn, 1931.

Weber, David. *The Spanish Frontier in North America.* New Haven, CT: Yale University Press, 1992.

Webster, Noah. *An American Dictionary of the English Language.* New York: S. Converse, 1828. Reprint, San Francisco: Foundation for American Christian Education, 1996.

Wedel, Waldo. *Prehistoric Man on the Great Plains.* Norman: University of Oklahoma Press, 1961.

West, Elliott. *The Way to the West: Essays on the Central Plains.* Albuquerque: University of New Mexico Press, 1995.

White, David A., ed. *News of the Plains and Rockies, 1803–1865.* Vol. 3. Spokane, WA: Arthur H. Clark, 1997.

White, Richard. *It's Your Misfortune and None of My Own: A New History of the American West.* Norman: University of Oklahoma Press, 1991.

Winfrey, Dorman H., and James M. Day, eds. *The Indian Papers of Texas and the Southwest: 1825–1916.* Austin: Pemberton Press, 1966.

Wilkins, Thurman. *Cherokee Tragedy: The Story of the Ridge Family and the Decimation of a People.* New York: MacMillan, 1970.

Wood, Richard George. *Stephen Harriman Long, 1784–1864: Army Engineer, Explorer, Inventor.* Glendale, CA: Arthur H. Clark, 1966.

Wooster, Ralph A. *Lone Star: Blue and Gray.* Austin: Texas State Historical Association, 1995.

Worster, Donald. *Dust Bowl: The Southern Plains In the 1930s.* New York: Oxford University Press, 1979.

ARTICLES

Abel, Annie H. "Proposal for an Indian State." *Annual Report of the American Historical Association* 1: 89–104. Washington, DC: GPO, 1908.

Alford, Terry L. "The West as a Desert in American Thought Prior to Long's 1819–1820 Expedition." *Journal of the West* 8 (October 1969): 515–25.

Allen, John C. "The Garden-Desert Continuum: Competing Views of the Great Plains in the Nineteenth Century." *Great Plains Quarterly* 5 (Fall 1985): 209.

Ashcraft, Allan C. "Confederate Indian Department Conditions in August, 1864." *The Chronicles of Oklahoma* 41, no. 3 (Fall 1963): 271–74.

———. "Confederate Indian Territory Conditions in 1865." *Chronicles of Oklahoma* 42, no. 4 (Winter 1964–65): 421–28.

Banks, Dean. "Civil War Refugees from Indian Territory in the North, 1861–1864." *Chronicles of Oklahoma* 41 (Fall 1963): 286–98.

Barnes, Lela, ed. "Journal of Isaac McCoy for the Exploring Expedition of 1828." *Kansas Historical Quarterly* 5, no. 3 (August 1936): 227–77.

———. "Journal of Isaac McCoy for the Exploring Expedition of 1830." *Kansas Historical Quarterly* 5, no. 4 (November 1936): 339–77.

Bears, Edwin C. "The Civil War Comes to Indian Territory, 1861: The Flight of Opothleyoholo." *Journal for the West* 11 (1972): 9–42.

Bender, Averam. "Frontier Defense in the Territory of New Mexico, 1846–1853." *New Mexico Historical Review* 9 (1934): 249–72.

———. "The Texas Frontier, 1848–1861." *Southwest Historical Quarterly* 38 (October 1934): 135–48.

Blasing, T. J., D. W. Stahle, and D. N. Duvick. "Tree Ring–Based Reconstruction of Annual Precipitation in the South-Central United States from 1750–1980." *Water Resources Research* 24 (January 1988): 163–71.

Bonnifield, Paul. "The Choctaw Nation on the Eve of the Civil War." *Journal of the West* 12 (July 1973): 386–402.

Bowden, Martyn J. "The Perception of the Western Interior of the United States, 1800–1870: A Problem in Historical Geosophy." *Proceedings of the Association of American Geographers* 1 (1969): 16–21.

――――. "Desertification of the Great Plains: Will It Happen?" *Economic Geography* 53 (1977): 397–406.

Brooks, Robert L., and Robert Bell. "The Last Prehistoric Villagers." *Chronicles of Oklahoma* 67 (Fall 1989): 296–318.

Brown, William R. "Natural History of the Canadian River, 1820–1853." *Panhandle Plains Historical Review* 61 (1988): 1–15.

Brown, Kenny L. "Building a Life: Culture, Society, and Leisure in the Cherokee Outlet." *Chronicles of Oklahoma* 71 (Summer 1993): 174–201.

Bullard, Clara W. "Pioneer Days in the Cherokee Strip." *Chronicles of Oklahoma* 36 (Fall 1958): 258–69.

Caldwell, Martha B. "Some Kansas Rainmakers." *Kansas Historical Quarterly* 7 (August 1938): 306–24.

Carroll, H. Bailey, ed. "The Journal of Lieutenant J. W. Abert from Bent's Fort to Saint Louis in 1845," *Panhandle-Plains Historical Review* 14 (1941): 3–113.

Carroll, Lew F. "An Eighty-Niner Who Pioneered the Cherokee Strip." *Chronicles of Oklahoma* 24 (Spring 1946): 87–101.

Carson, Gerald. "The Rainmakers." *Natural History* (September 1985): 20–28.

Carson, William G. B., ed. "William Carr Lane Diary." *New Mexico Historical Review* 39 (1964): 181–234.

Chapman, Berlin B. "Opening of the Cherokee Outlet: An Archival Study." *Chronicles of Oklahoma* 40 (Summer, Fall 1962): 158–81, 253–85.

――――. "The Final Report of the Cherokee Commission." *Chronicles of Oklahoma* 19 (December 1941): 356–67.

Cook, Edward, David Meko, David Stahle, and Malcolm Cleaveland. "Drought Reconstructions for the Continental United States." *American Meteorological Society* 12 (April 1999): 1145–62.

Corbett, William. "Peerless Princess of the Best Country: The Early Years of Tonkawa." *Chronicles of Oklahoma* 62 (Winter 1984–85): 388–407.

Crimmins, M. L., ed. "Robert E. Lee in Texas: Letters and Diary." *West Texas Historical Association Yearbook* 8 (June 1932): 16–24.

Davis, Athie. "Reminiscences of Pioneer Days in Garfield County." *Chronicles of Oklahoma* 35 (Summer 1957): 163–68.

Debo, Angie. "Southern Refugees of the Cherokee Nation." *Southwestern Historical Quarterly* 35, no. 4 (April 1932): 255–266.

Dillehay, Tom D. "Late Quaternary Bison Population Changes on the Southern Plains." *Plains Anthropologist* 19 (1974): 180–96.

Dillon, Richard. "Stephen Long's Great American Desert." *Proceedings of the American Philosophical Society* 111 (April 1967): 93–108.

Doran, Michael F. "Antebellum Cattle Herding in the Indian Territory." *Geographical Review* 66 (January 1976): 48–58.

Droze, W. H. "Changing the Plains Environment: The Afforestation of the Trans-Mississippi West." *Agricultural History* 51 (January 1977): 6–22.

Edwards, John. "An Account of My Escape from the South in 1861." *Chronicles of Oklahoma* 43 (Spring 1965): 58–89.

Edwards, John. "The Choctaw Indians in the Middle of the Nineteenth Century." *Chronicles of Oklahoma* 10 (1965): 392–425.

Estep, Raymond. "Lieutenant William E. Burnet Letters: Removal of the Texas Indians and the Founding of Fort Cobb." *Chronicles of Oklahoma* 38 (1960) no. 3, 274–309; no. 4, 369–96.

Everett, Edward. "Long's Expedition." *North American Review* 16 (April 1823): 242–69.

Fisher, LeRoy H. "A Civil War Experience of Some Arkansas Women in Indian Territory." *Chronicles of Oklahoma* 57, no. 2 (Summer 1979): 137–63.

Fisher, LeRoy H., and William L. McMurry. "Confederate Refugees from Indian Territory." *Chronicles of Oklahoma* 57 (Winter 1979): 451–62.

Flores, Dan. "Bison Ecology and Bison Diplomacy: The Southern Plains from 1800–1850." *Journal of American History* 78 (September 1991): 465–85.

Gage, Duane. "Oklahoma: A Resettlement Area for Indians." *Chronicles of Oklahoma* 47 (Fall 1969): 282–97.

Gibson, Arrell Morgan. "Ranching on the Southern Great Plains." *Journal of the West* 6 (January 1967): 135–53.

Gilstrap, Lucille. "Homesteading the Strip." *Chronicles of Oklahoma* 51 (Fall 1973): 285–304.

Graebner, Norman Arthur. "Pioneer Indian Agriculture in Oklahoma." *Chronicles of Oklahoma* 23 (Summer, Fall 1945): 232–48.

Harper, H. "Drought in Central Oklahoma from 1710–1959, Calculated from Annual Rings of Post Oak Trees." *Proceedings of the Oklahoma Academy of Science* 41 (1961): 23–29.

Harrington, Mark. "Weather Making, Ancient and Modern." *National Geographic* 6 (April 1894): 35–62.

Heath, Gary. "The First Federal Invasion of Indian Territory." *Chronicles of Oklahoma* 44, no. 4 (Winter 1966–67): 409–19.

Hewitt, George B. *Review of Factors Affecting Fecundity, Oviposition, and Egg Survival of Grasshoppers in North America.* [Beltsville, MD]: U.S. Department of Agriculture, Agriculture Research Service 1985.

Hicks, Hannah. "The Diary of Hannah Hicks." *The American Scene* 13 (1972): n.p.

Holden, W. C. "West Texas Drouths." *The Southwestern Historical Quarterly* 32 (October 1928): 103–23.

———. "Frontier Defense, 1846–1860." *West Texas Historical Association Yearbook* 6 (June 1930): 35–64.

Johnson, B. H. "Singing Oklahoma's Praises: Boosterism in the Sooner Land." *Great Plains Journal* 11 (1971): 57–65.

Kennan, Clara B. "Neighbors in the Cherokee Strip." *Chronicles of Oklahoma* 27 (Spring 1949): 76–88.

Klos, George. "'Our People Could not Distinguish One Tribe from Another': The 1859 Expulsion of the Reserve Indians from Texas." *Southwestern Historical Quarterly* 97 (April 1994): 599–619.

Kreidler, Tai. "To Anchor the Wind." *Journal of the West* 29, vol. 4 (1990): 46–52.

Lawson, Merlin P., and Charles W. Stockton. "Desert Myth and Climate Reality." *Annals of the Association of American Geographers* 71 (December 1981): 527–35.

Lewis, Anna, ed. "Diary of a Missionary to the Choctaws, 1860–1861." *Chronicles of Oklahoma* 17 (1939): 428–47.

"Major Long's Second Expedition." *North American Review* 21 (July–December, 1825): 178–89.

Malin, James C. "Dust Storms: Part 1, 1850–1860." *Kansas Historical Quarterly* 14 (May 1946): 129–44.

———. "Dust Storms: Part 2, 1861–1880." *Kansas Historical Quarterly* 14 (May 1946): 265–96.

———. "Dust Storms: Part 3, 1881–1900." *Kansas Historical Quarterly* 14 (May 1946): 391–413.

McDermott, John Francis. "Isaac McCoy's Second Exploring Trip in 1828." *Kansas Historical Quarterly* 13 (February 1945): 400–62.

McLoughlin, William G. "The Choctaw Slave Burning: A Crisis in Mission Work among the Indians." *Journal of the West* 13 (October 1974): 113–27.

———. "Indian Slaveholders and Presbyterian Missionaries, 1837–1861." *Church History* 42 (December 1973): 535–51.

Milam, Joe B. "The Opening of the Cherokee Outlet." *Chronicles of Oklahoma* 9 (1931): Part 1 (September), 269–86; Part 2 (December), 454–75.

Miner, Craig. "Cherokee Sovereignty in the Gilded Age: The Outlet Question." *Chronicles of Oklahoma* 71 (Summer 1993): 118–37.

Morgan, James F. "Choctaw Warrants of 1863." *Chronicles of Oklahoma* 57 (Spring 1979): 55–66.

Morris, Ralph C. "The Notion of a Great American Desert East of the Rockies." *Mississippi Valley Historical Review* 13, no. 2 (1927): 190–200.

Morrison, James D. "Note on Abolitionism in the Choctaw Nation." *Chronicles of Oklahoma* 38 (Spring 1960): 78–83.

Morton, Orland. "Confederate Government Relations with the Five Civilized Tribes, Part 2." *Chronicles of Oklahoma* 31, no. 3 (Autumn 1953): 299–323.

Muhs, Daniel, and Vance Holliday. "Evidence of Active Dune Sand on the Great Plains in the Nineteenth Century from Accounts of Early Explorers." *Quaternary Research* 43 (1995): 198–208.

Neighbours, Kenneth. "Robert Simpson Neighbors and the Founding of the Texas Indian Reservations." *West Texas Historical Association Yearbook* 31 (1955): 65–74.

———. "Chapters from the History of Texas Indian Reservations." *West Texas Historical Association Yearbook* 33 (1957): 3–16.

———. "The Assassination of Robert S. Neighbors." *West Texas Historical Association Yearbook* 34 (1958): 38–49.

———. "Indian Exodus out of Texas in 1859." *West Texas Historical Association Yearbook* 36 (1960): 80–97.

Niles, Hezekiah, ed. "American Desert." *Niles' Weekly Register* 35 (September 1828): 70.

Oliva, Leo E. "Fort Atkinson on the Santa Fe Trail, 1850–1854." *Kansas Historical Quarterly* 40 (Summer 1974): 212–33.

Peale, Titian R. "Journal of Titian Ramsay Peale, Pioneer Naturalist." Edited by A. O. Weese. *Missouri Historical Review* 41 (January 1947): 147–63.

Prucha, Francis Paul. "Indian Removal and the Great American Desert." *Indiana Magazine of History* 59 (December 1963): 299–322.

"Removal of the Indians." *North American Review* 30, no. 66 (1830): 62–121.

Richardson, Rupert N. "The Comanche Reservation in Texas." *West Texas Historical Association Yearbook* 5 (1929): 47–71.

Rister, Carl Coke. "Free Land Hunters of the Southern Plains." *Chronicles of Oklahoma* 22 (Winter 1944–1945): 392–401.

———. "'Oklahoma,' the Land of Promise." *Chronicles of Oklahoma* 23 (Spring 1945): 2–15.

Sageser, A. Bower. "Editor Bristow and the Great Plains Irrigation Revival of the 1890s." *Journal of the West* 3 (January 1964): 75–89.

Savage, William W., Jr. "Of Cattle and Corporations: The Rise, Progress, and Termination of the Cherokee Strip Livestock Association." *Chronicles of Oklahoma* 71, no. 2 (Summer 1993): 138–53.

Schwartz, Lillian Carlile. "Life in the Cherokee Strip." *Chronicles of Oklahoma* 42 (Summer 1964): 62–74.

Shaw, James H., and Martin Lee. "Relative Abundance of Bison, Elk and Pronghorn on the Southern Plains, 1806–1857." *Plains Anthropologist* 42 (February 1997): 163–72.

Silliman, Benjamin. "Expedition of Major Long and Party, to the Rocky Mountains." *American Journal of Science and Arts* 6 (1823): 178–89.

Smith, Henry Nash. "Rain Follows the Plow: The Notion of Increased Rainfall for the Great Plains, 1844–1880." *Huntington Library Quarterly* 10, no. 2 (1946–1947): 169–93.

Smith, Robert E. "The Wyandot Exploring Expedition of 1839." *Chronicles of Oklahoma* 55 (Fall 1977): 282–92.

Spence, Clark C. "The Dyrenforth Rainmaking Experiments: A Government Venture in Pluviculture." *Journal of the Southwest* 3, no. 3 (Fall 1961): 205–32.

Stockton, Charles, and David Meko. "Drought Recurrence in the Great Plains as Reconstructed from Long-Term Tree-Ring Records." *American Meteorological Society* 23 (January 1983): 17–29.

Tucker, John M. "Major Long's Route from the Arkansas to the Canadian River, 1820." *New Mexico Historical Review* 38 (July 1963): 184–219.

Turk, Eleanor L. "Selling the Heartland: Agents, Agencies, Press, and Policies Promoting German Emigration to Kansas in the Nineteenth Century." *Kansas History* 12 (Fall 1989): 150–59.

Turner, Alvin O. "Order and Disorder: The Opening of the Cherokee Outlet." *Chronicles of Oklahoma* 71, no. 2 (Summer 1993): 154–73.

Webb, Walter P. "Some Vagaries of the Search for Water in the Great Plains." *Panhandle-Plains Historical Review* 3 (1930): 28–37.

Wessel, Thomas R. "Prologue to the Shelterbelt, 1870 to 1934." *Journal of the West* 6 (January 1967): 119–34.

Trickett, Dean. "The Civil War in Indian Territory, 1861." *Chronicles of Oklahoma* 17, no. 3 (September–December 1939): 315–27.

———. "The Civil War in Indian Territory, 1861 (Continued)." *Chronicles of Oklahoma* 17, no. 4 (June 1940): 401–12.

———. "The Civil War in Indian Territory, 1861 (Continued)." *Chronicles of Oklahoma* 18, no. 3 (September 1940): 266–79.

———. "The Civil War in Indian Territory, 1861 (Continued)." *Chronicles of Oklahoma* 19, no. 1 (March 1941): 55–69.

———. "The Civil War in Indian Territory, 1861 (Continued)." *Chronicles of Oklahoma* 19, no. 4 (December 1941): 381–96.

Turner, Alvin O. "Order and Disorder: The Opening of the Cherokee Outlet." *Chronicles of Oklahoma* 71 (Summer 1993): 154–73.

Wade, J. S. "Uncle Sam's Horse Race." *Chronicles of Oklahoma* 35 (Summer 1957): 147–53.

White, William H. "The Texas Slave Insurrection of 1860." *Southwest Historical Quarterly* 52 (January 1949): 259–85.

Williams, J. W. "A Statistical Study of the Drouth of 1886." *West Texas Historical Association Yearbook* 21 (October 1945): 85–109.

Willey, William J. "The Second Federal Invasion of Indian Territory." *Chronicles of Oklahoma* 44, no. 4 (Winter 1966–1967): 420–43.

Wilson, Paul T. "Delegates of the Five Civilized Tribes to the Confederate Congress." *Chronicles of Oklahoma* 53 (1975): 353–66.

Windham, William T. "The Problem of Supply in the Trans-Mississippi Confederacy." *The Journal of Southern History* 27, no. 2 (May 1961): 149–68.

Wood, Richard G. "Dr. Edwin James, a Disappointed Explorer." *Minnesota History* 34 (Fall 1955): 284–86.

Woodhouse, Connie, and Jonathan Overpeck. "2000 Years of Drought Variability in the Central United States." *Bulletin of the American Meteorological Society* 79 (December 1998): 2693–2714.

DISSERTATION AND THESES

Buenger, Walter Louis. "Stilling the Voice of Reason: Texas and the Union, 1854–1861." PhD dissertation, Rice University, 1979.

Jones, Andrew McKee. "Comanches and Texans in the Making of the Comanche Nation: The Historical Anthropology of Comanche-Texan Relations, 1803–1997." PhD dissertation, University of Wisconsin–Madison, 1997.

Lynn-Sherow, Bonnie. "Ordering the Elements: An Environmental History of West Central Oklahoma." PhD dissertation, Northwestern University, 1998.

Neighbors, Alice Atkinson. "Life and Public Works of Robert S. Neighbors." Master's thesis, University of Texas, 1936.

Index

References to illustrations appear in italics.